Materials for Engineering

Second edition

W. Bolton

Newnes

OXFORD AUCKLAND BOSTON JOHANNESBURGH MELBOURNE NEW DELHII

Newnes
An imprint of Butterworth-Heinemann
Linacre House, Jordan Hill, Oxford OX2 8DP
225 Wildwood Avenue, Woburn, MA 01801-2041
A division of Reed Educational and Professional Publishing Ltd

A member of the Reed Elsevier plc group

First published 1994
Second edition 2000

British Library Cataloguing in Publication Data
A catalogue record for this book is available from the British Library

ISBN

Library of Congress Cataloguing in Publication Data
A catalogue record for this book is available from the Library of Congress

Printed and bound in Great Britain

Contents

Preface

Aims

The aims of this book are to introduce engineering students to:

- The identification and classification of commonly used engineering materials as – metals, ferrous and non-ferrous; polymers, thermosetting, thermoplastics and elastomers; ceramics; composites
- The properties of materials, the obtaining of data on properties and the tests to obtain the properties
- How the properties of materials affect the applications for which they can be used and the processing methods that can be used
- The recognition of how the properties may be changed through modifications in composition, structure and processing
- The selection of materials and processes for particular applications

Changes from first edition

This second edition differs from the first edition in having:

- More discussion of materials
- More data tables given for materials
- More details of processing and the selection of processes
- More case studies
- Student activities included

Target courses

The second edition of this book has been designed to comprehensively cover the latest versions of the General National Vocational Qualifications (GNVQ) unit at Advanced level of *Engineering Materials*, this being a mandatory unit in Engineering vocational qualifications at this level. GNVQs are available in schools and colleges throughout England, Wales and Northern Ireland; they have the main aim of raising the status of vocational education within a new system of high quality vocational qualifications. GNVQs provide a broad-based vocational education that continues many aspects of secondary education.

W. Bolton

1 Materials

1.1 Classification of materials

The term *engineering* material is used for a material designed, made and used for a practical purpose, e.g. carrying a mechanical load or perhaps providing a path for an electric current. You are surrounded by products involving a wide range of such materials, e.g. the motorcar with its shaped metal exterior surface, an electric cable with its metal core to carry the current and plastic sheath for electric insulation, the mobile phone with its plastic casing, a house with its brick structure and glass windows, a desk made of wood.

There are many materials in use and it is convenient to classify them into four main groups, these being determined by their internal structure and consequential properties. In everyday use, we identify materials in these groups and distinguish between them by their characteristic properties rather than the internal structure responsible for the properties. Thus we might look at a particular material and handle it and come to the conclusion that it is easily bent (the term used is ductile), cold to touch (this means a high thermal conductivity since it readily conducts the heat away from you hand), not easily broken, etc.

1 *Metals*

Essentially, metals are based on metallic chemical elements and, in general, have high electrical conductivities, high thermal conductivities, can be ductile and thus permit products to be made by being bent into shape, and have a relatively high stiffness and strength. Thus the bodywork of a car is generally made of metal which is bent into the required shape and has reasonable stiffness and strength so that it retains that shape. The core of an electric cable is made of a metal to give the high electrical conductivity required.

Engineering metals are generally alloys, *alloys* being metallic materials formed by mixing two or more elements. For example, mild steel is an alloy of iron and carbon, stainless steel an alloy of iron, chromium, carbon, manganese and possibly other elements. The reason for adding elements to the iron is to improve the properties since pure metals are very weak materials. The carbon improves the strength of the iron, the chromium in the stainless steel improves the corrosion resistance.

Metals are also classified as follows:

(a) *Ferrous alloys*

These are iron-based alloys, e.g. steels and cast irons.

(b) *Non-ferrous alloys*
These are not iron-based alloys, e.g. aluminium-based and copper-based alloys.

2 *Polymers (plastics)*
Essentially, polymers are based on large organic molecules, i.e. carbon containing molecules. They can be classified as:

(a) *Thermoplastics*
Thermoplastics soften when heated and become hard again when the heat is removed. The term implies that the material becomes 'plastic' when heat is applied. Thus thermoplastic materials can be heated and bent to form required shapes, thermosets cannot. Thermoplastic materials are generally flexible and relatively soft. Polythene is an example of a thermoplastic, being widely used as films or sheets for such items as bags, 'squeezy' bottles, and wire and cable insulation.

(b) *Thermosets*
Thermosets do not soften when heated, but char and decompose. Thermosets are rigid and hard. Phenol formaldehyde, known as Bakelite, is a thermoset. It is widely used for electrical plug casings, door knobs and handles.

The term *elastomers* is used for polymers which by their structure allow considerable extensions that are reversible, e.g. rubber bands.

In general, polymers have low electrical conductivity and low thermal conductivity, hence their use for electrical and thermal insulation. Compared with metals, they have lower densities, expand more when there is a change in temperature, are generally more corrosion resistant, have a lower stiffness, stretch more and are not as hard. When loaded they tend to creep, i.e. the extension gradually changes with time. Their properties depend very much on the temperature so that a polymer which may be tough and flexible at room temperature may be brittle at 0°C and show considerable creep at 100°C.

3 *Ceramics*
Ceramics are based on inorganic compounds and were originally just clay-based materials. They tend to be brittle, relatively stiff, stronger in compression than tension, hard, chemically inert and bad conductors of electricity and heat. The non-glasses tend to have good heat and wear resistance and high-temperature strength. Ceramics include:

(a) *Glasses*: soda line glasses, borosilicate glasses, pyroceramics

(b) *Domestic ceramics*: porcelain, vitreous china, earthenware, stoneware, cement
Examples of domestic ceramics and glasses abound in the home in the form of cups, plates and glasses.

(c) *Engineering ceramics*: alumina, carbides, nitrides
Because of their hardness and abrasion resistance, such ceramics are widely used as the cutting edges of tools.

(d) *Natural ceramics*: rocks

4 *Composites*
Composites are materials composed of two different materials bonded together. For example, there are composites involving glass fibres or particles in polymers, ceramic particles in metals (cermets), and steel rods in concrete (reinforced concrete). Wood is a natural composite consisting of tubes of cellulose in a polymer called lignin. Composites made with fibres embedded, all aligned in the same direction in some matrix, will have properties in that direction markedly different from properties in other directions. Composites can be designed to combine the good properties of different types of materials while avoiding some of their drawbacks.

If the above classification was solely in terms of properties it would be deemed rather crude. Within each group there is a large variation in properties and there are no clear property boundaries between them.

1.2 Examples of material identification

The following are examples of material identification based on their properties:

1 *The envelope of an electric light bulb*
This is transparent, rigid, brittle, bad conductor of electricity, and able to withstand getting reasonably hot without deforming or melting. The material is a glass. A transparent plastic would not be likely to be able to get reasonably hot without deforming and is less likely to be so brittle. The light bulb might, however, be mounted inside a decorative casing which is made of plastic.

2 *A screwdriver*
The blade of the screwdriver is stiff, hard and not easily broken; it is made of a metal (a high-carbon steel). The handle of the screwdriver is not hard but still relatively stiff; it is made usually of a plastic since this material is easily formed into the required shape and can have sufficient stiffness not to deform noticeably when the handle is turned.

3 *Spark plug of an internal combustion engine*
The spark electrodes have to conduct electricity and so have to be made of a metal (a tungsten alloy because it has to be highly resistance to the wear effects produced by the spark and chemical attack from the hot gases in the engine cylinder). The central electrode has to be electrically insulated from the outer electrode at high temperatures and thus a ceramic (alumina) is used.

Activity
List ten common products and identify the types of materials present.

Activity

Examine the bodywork of a car and identify the type of material present, giving reasons for your answer

1.3 The evolution of materials

The early history of the human race can be divided into periods according to the materials that were predominantly used. Thus we have the Stone Age, the Bronze Age and the Iron Age.

In the *Stone Age* (about 8000 BC to 4000 BC), people could only use the materials they found around them such as stone, wood, clay, animal hides, bone, etc. The tools they made were limited to what they could fashion out of these materials. Thus tools were limited to those that could be made from stone, flint, bone and horn.

By about 4000 BC, people in the Middle East were able to extract *copper* from its ore and it rapidly became an important material. Copper is a ductile material which can be hammered into shapes, thus enabling a greater variety of items to be fashioned than was possible with stone. Because the copper ores contained impurities that were not completely removed by the smelting, alloys were produced. It was soon realised that the deliberate adding of additives to copper could produce materials with improved properties. About 2000 BC it was found that when tin was added to copper, an alloy was produced that had an attractive colour, was easy to form and harder than copper alone. This alloy was called *bronze*. Thus we have the *Bronze Age*.

About 1200 BC the extraction of *iron* from its ores signalled another major development, hence the *Iron Age*. Iron in its pure form was, however, inferior to bronze but by heating items fashioned from iron in charcoal and hammering them, a tougher material, called *steel*, was produced. Plunging the hot metal into cold water, i.e. quenching, was found to improve the hardness. It was also found that reheating and cooling the metal slowly produced a less hard but tougher and less brittle material, this process now being termed tempering. Thus *heat treatment processes* were developed.

The large-scale production of iron can be considered an important development in the evolution of materials in that it made the material more widely and cheaply available for products. Large-scale iron production with the first coke-fuelled blast furnace started in 1709. Cast iron was used in 1777 to build a bridge at the place in England now known as Ironbridge. The term *industrial revolution* is used for the period that followed as the pace of developments of materials and machines increased rapidly and resulted in major changes in the industrial environment and the products generally available. The year 1860 saw the development of the Bessemer and open hearth processes for the production of steel, and this date may be considered to mark the general use of steel as a constructional material. Aluminium was extracted from its ores in 1845 and produced commercially in 1886. In the years that followed, many new alloys were developed. The high strength aluminium alloy Duralumin was developed in 1909, stainless steel in 1913, high strength nickel–chromium alloys for high temperature use in 1931. Titanium was first produced commercially in 1948.

While naturally occurring *plastics* have been used for many years, the first manufactured plastic, celluloid, was not developed until 1862. In 1906 Bakelite was developed. The period after about 1930, often termed the *Plastic Age*, saw a major development of plastics and their use in a wide range of products. Polyethylene is an example of a scientific investigation yielding a surprise result. In 1931 a Dutch scientist, A. Michels, was given approval by ICI to design apparatus that could be used to carry out research into the effects of high pressure on chemical reactions. In 1932 the investigation began and in March 1933 a surprise result was obtained. The chemical reaction between ethylene and benzaldehyde was being studied at a pressure of 2000 times atmospheric pressure and a temperature of 170°C when a waxy solid was found to form. The material that had been formed was *polyethylene*. The experiment had not been designed to develop a new material but that was the outcome. The commercial production of polyethylene started in England in 1941. The development of *polyvinyl chloride* was, however, an investigation where a new material was sought. In 1936 there was no readily available material that could replace natural rubber. In the event of a war Britain's natural rubber supply would be at risk since it had to travel by sea from the Far East. Thus a substitute was required and research was initiated. In July 1940 a small amount of PVC was produced, but there was to be many problems before commercial production of PVC with suitable properties could begin in 1945 .

The evolution of materials over the years has resulted in changes in our lifestyles. Thus when tools were limited to those that could be fashioned out of stone, there was severe limitations on what could be achieved with them. The development of metals enabled finer products to be fashioned. For example, bronze swords were far superior weapons to stone weapons. The development of plastics has enabled a great range of products to be produced cheaply and in large numbers: consider what the world would be like today if plastics had not been developed.

Activity

Find how the materials and consequent structure of cars has changed over the years. A key date is 1923 when the American Rolling Mill Company opened the first continuous hot-strip mill for rolling thin steel.

2 Properties of materials

2.1 Introduction to materials selection

What materials could be used for containers of Coca-Cola? Well you can buy Coca-Cola in aluminium cans, in glass bottles and in plastic bottles. What makes these materials suitable, and others not? In order to attempt to answer this question we need to discuss the properties of materials.

Thus we might talk about the need for the container material to be:

1. Rigid, so that the container does not become stretched unduly, i.e. become floppy, under the weight of the Coca-Cola.
2. Strong, so that the container can stand the weight of the Coca-Cola without breaking.
3. Resistant to chemical attack by the Coca-Cola.
4. Able to keep the 'fizz' in the Coca-Cola, i.e. not to allow the gas to escape through the walls of the container.
5. Low density so that the container is not too heavy.
6. Cheap.
7. Easy and cheap to process to produce the required shape.

You can no doubt think of more requirements. The selection of a material thus involves balancing a number of different requirements and making a choice of the material which fulfils as many as possible as well as possible.

Consider another product, a bridge. What are the requirements for the material to be used in a bridge? These are likely to include:

1. Strength so that when the bridge is subject to loads, such as people, cars, lorries, etc. crossing it, it will not break.
2. Stiff enough so that the bridge will not stretch unduly under the load.
3. Can be produced and joined in long enough lengths to be able to span the gap to be bridged.
4. The materials costs and the fabrication costs are not too high.
5. Resists or can be protected from atmospheric corrosion.
6. Can be maintained at a reasonable cost over a period of years.

You can no doubt add more requirements. Materials which are used are wood, steel and reinforced concrete.

The selection of a material depends on the properties required of it in order that it can fulfil the uses required of it.

2.1.1 The requirements of materials

The selection of a material from which a product can be manufactured depends on a number of factors. These are often grouped under three main headings, namely:

1 The requirements imposed by the conditions under which the product is used, i.e. the service requirements. Thus if a product is to be subject to forces then it might need strength, if subject to a corrosive environment then it might require corrosive resistance.

2 The requirements imposed by the methods proposed for the manufacture of the product. For example, if a material has to be bent as part of its processing, the material must be ductile enough to be bent without breaking. A brittle material could not be used.

3 Cost.

2.2 Properties of materials

Materials selection for a product is based upon a consideration of the properties required. These include:

1 *Mechanical properties*
These are the properties displayed when a force is applied to a material and include strength, stiffness, hardness, toughness, and ductility.

2 *Electrical properties*
These are the properties displayed when the material is used in electrical circuits or components and include resistivity, conductivity, and resistance to electrical breakdown.

3 *Magnetic properties*
These are relevant when the material is used as, for example, a magnet or part of an electrical component such as an inductor which relies on such properties.

3 *Thermal properties*
These are the properties displayed when there is a heat input to a material and include expansivity and heat capacity.

4 *Physical properties*
These are the properties which are characteristic of a material and determined by its nature, e.g. density, colour and surface texture.

5 *Durability properties*
These are properties such as stability and resistance to environmental degradation, corrosion, chemical solubility and absorption.

In discussing the properties of materials it is important to recognise that they are often markedly changed by the temperature at which they are used and any treatments the materials undergo. For example, a plastic may be relatively stiff at room temperature but far from stiff at the boiling point of water. A steel may be ductile at 20°C but become brittle at temperatures below −10°C. Steels can have their properties changed by heat treatment, such as *annealing* which involves heating to some temperature and slowly cooling. This renders the material soft and ductile. Heating a steel to some temperature and then *quenching*, i.e. immersing the hot material in cold water, can be used to make a steel harder, stronger and less ductile. Materials can also have their properties changed by working. For example, if you take a piece of carbon steel and permanently deform it, perhaps by bending it, then it will have different mechanical properties to those existing before that deformation. It is said to be *work hardened.*

In the following, some of the properties listed above are discussed and the quantities which are used as a measure of them defined. Appendix B lists these, together with other properties and terms. Appendix A is a discussion of units and unit prefixes.

2.3 Mechanical properties

The mechanical properties are about the behaviour of materials when subject to forces. When a material is subject to external forces, then internal forces are set up in the material which oppose the external forces. The material can be considered to be rather like a spring. A spring, when stretched by external forces, sets up internal opposing forces which are readily apparent when the spring is released and they force it to contract. A material subject to external forces which stretch it is said to be in *tension* (Figure 2.1(a)). A material subject to forces which squeeze it is said to be in *compression* (Figure 2.1(b)). If a material is subject to forces which cause it to twist or one face slide relative to an opposite face then it is said to be in *shear* (Figure 2.1(c)). An object, in some situations, can be subject to both tension and compression, e.g. a beam (Figure 2.2) which is being bent, the bending causing the upper surface to contract and so be in compression and the lower surface to extend and be in tension.

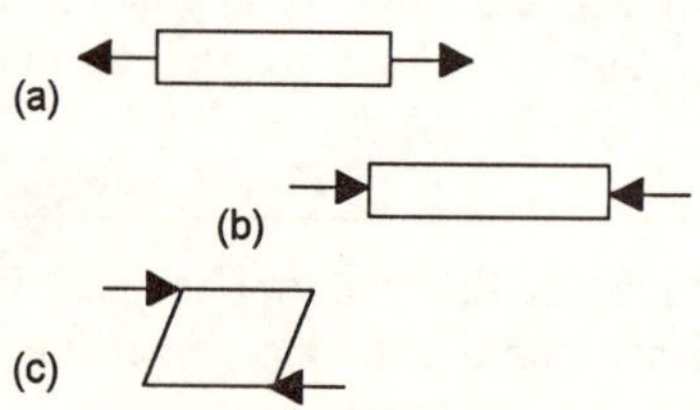

Figure 2.1 *(a) Tension, (b) compression, (c) shear*

2.3.1 Stress and strain

In discussing the application of forces to materials an important aspect is often not so much the size of the force as the force applied per unit area. Thus, for example, if we stretch a strip of material by a force F applied over its cross-sectional area A (Figure 2.3), then the force applied per unit area is F/A. The term *stress*, symbol σ, is used for the force per unit area:

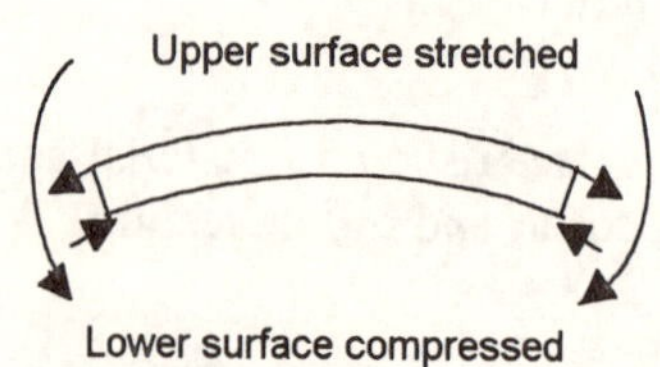

Figure 2.2 *Bending*

$$\text{stress} = \frac{\text{force}}{\text{area}}$$

Stress has the units of pascal (Pa), with 1 Pa being a force of 1 newton per square metre, i.e. 1 Pa = 1 N/m^2 (see Appendix A on units). The

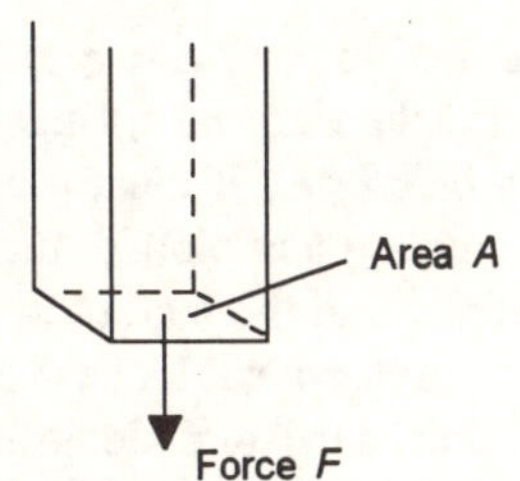

Figure 2.3 *Stress*

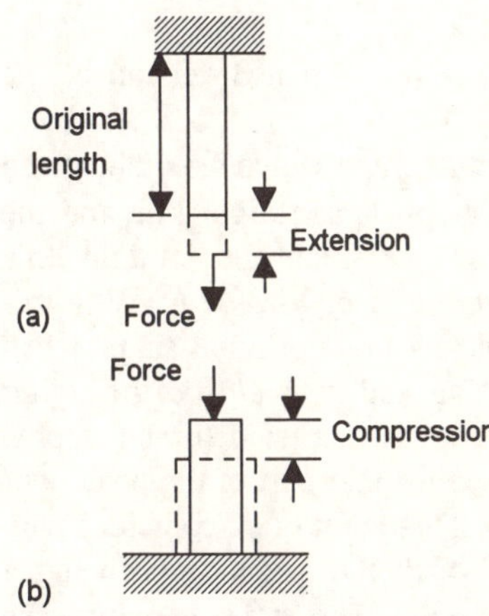

Figure 2.4 *(a) Tensile strain, (b) compressive strain*

stress is said to be *direct stress* when the area being stressed is at right angles to the line of action of the external forces, as when the material is in tension or compression. Shear stresses are not direct stresses since the forces being applied are in the same plane as the area being stressed. The area used in calculations of the stress is generally the original area that existed before the application of the forces. The stress is thus sometimes referred to as the *engineering stress*, the term *true stress* being used for the force divided by the actual area existing in the stressed state.

When a material is subject to tensile or compressive forces, it changes in length (Figure 2.4). The term *strain*, symbol ε, is used for:

$$\text{strain} = \frac{\text{change in length}}{\text{original length}}$$

Since strain is a ratio of two lengths it has no units. Thus we might, for example, have a strain of 0.01. This would indicate that the change in length is 0.01× the original length. However, strain is frequently expressed as a percentage:

$$\text{Strain as a \%} = \frac{\text{change in length}}{\text{original length}} \times 100\%$$

Thus the strain of 0.01 as a percentage is 1%, i.e. this is when the change in length is 1% of the original length.

Example

A bar of material with a cross-sectional area of 50 mm^2 is subject to tensile forces of 100 N. What is the tensile stress?

The tensile stress is the force divided by the area and is thus:

$$\text{tensile stress} = \frac{100}{50}\ \text{N/mm}^2$$

$$= \frac{100}{50 \times 10^{-6}}\ \text{N/m}^2 \text{ or Pa}$$

$$= 2\ \text{MPa}$$

Example

A strip of material has a length of 50 mm. When it is subject to tensile forces it increases in length by 0.020 mm. What is the strain?

Strain is the change in length divided by the original length and is thus:

$$\text{strain} = \frac{0.020}{50} = 0.000\,04$$

Expressed as a percentage, the strain is

$$\text{strain} = \frac{0.020}{50} \times 100 = 0.04\%$$

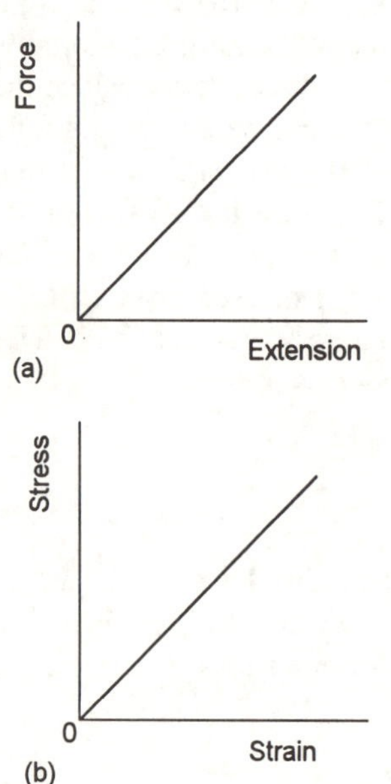

Figure 2.5 *Hooke's law*

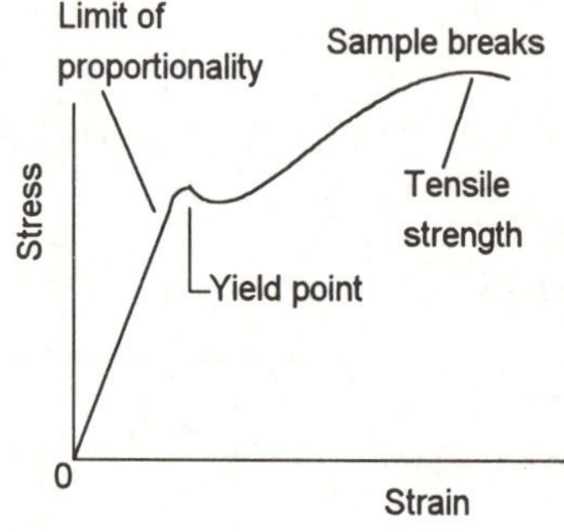

Figure 2.6 *Stress–strain graph for mild steel*

2.3.2 Stress-strain graphs

If gradually increasing tensile forces are applied to, say, a strip of mild steel then initially when the forces are released the material springs back to its original shape. The material is said to be *elastic*. If measurements are made of the extension at different forces and a graph plotted, then the extension is found to be proportional to the force and the material is said to obey *Hooke's law*. Figure 2.5(a) shows a graph when Hooke's law is obeyed. Such a graph applies to only one particular length and cross-sectional area of a particular material. We can make the graph more general so that it can be applied to other lengths and cross-sectional areas of the material by dividing the extension by the original length to give the strain and the force by the cross-sectional area to give the stress (Figure 2.5(b)). Then we have, for a material that obeys Hooke's law:

$$\text{stress} \propto \text{strain}$$

The stress-strain graph (Figure 2.5(b)) is just a scaled version of the force–extension graph in Figure 2.5(a).

Figure 2.6 shows the type of stress-strain graph which would be given by a sample of mild steel. Initially the graph is a straight line and the material obeys Hooke's law. The point at which the straight line behaviour is not followed is called the *limit of proportionality*. With low stresses the material springs back completely to its original shape when the stresses are removed, the material being said to be *elastic*. At higher forces this does not occur and the material is then said to show some *plastic* behaviour. The term plastic is used for that part of the behaviour which results in permanent deformation. This point often coincides with the point on a stress-strain graph at which the graph stops being a straight line, i.e. the *limit of proportionality*. The stress at which the material starts to behave in a non-elastic manner is called the *elastic limit*.

The *strength* of a material is the ability of it to resist the application of forces without breaking. The term *tensile strength* is used for the maximum value of the tensile stress that a material can withstand without breaking, i.e.

$$\text{tensile strength} = \frac{\text{maximum tensile forces}}{\text{original cross-sectional area}}$$

The *compressive strength* is the maximum compressive stress the material can withstand without becoming crushed. The unit of strength is that of stress and so is the pascal (Pa), with 1 Pa being 1 N/m^2. Strengths are often millions of pascals and so MPa is often used, 1 MPa being 10^6 Pa or 1 000 000 Pa.

With some materials, e.g. mild steel, there is a noticeable dip in the stress–strain graph at some stress beyond the elastic limit and the strain increases without any increase in load. The material is said to have yielded and the point at which this occurs is the *yield point*. For some

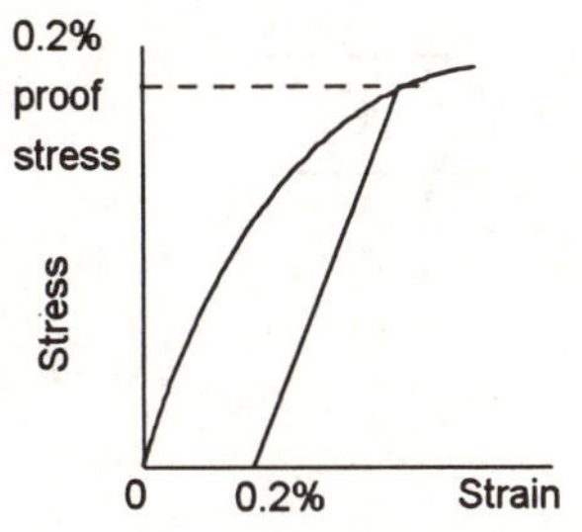

Figure 2.7 *0.2% proof stress*

materials, such as mild steel, there are two yield points termed the upper yield point and the lower yield point. A carbon steel typically might have a tensile strength of 600 MPa and a yield stress of 300 MPa.

Some materials, such as aluminium alloys (Figure 2.7), do not show a noticeable yield point and it is usual here to specify *proof stress*. The 0.2% proof stress is obtained by drawing a line parallel to the straight line part of the graph but starting at a strain of 0.2%. The point where this line cuts the stress–strain graph is termed the 0.2% yield stress. A similar line can be drawn for the 0.1% proof stress.

Figure 2.8 shows stress–strain graphs for a number of common materials. Stress-strain graphs are discussed in more detail in Chapter 3 when their determination and deductions that can be made from them are considered. Table 2.1 gives typical tensile strength values.

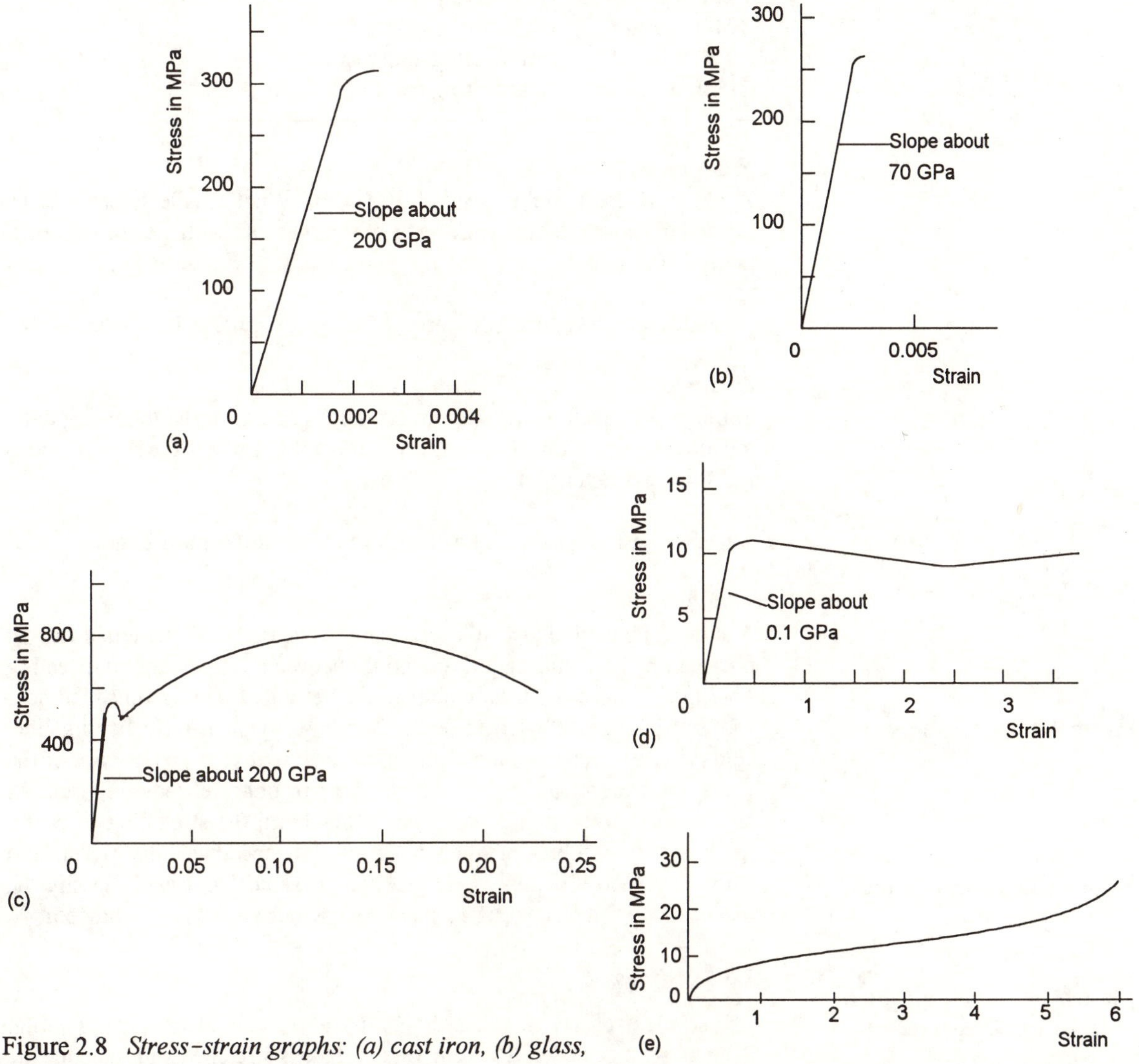

Figure 2.8 *Stress–strain graphs: (a) cast iron, (b) glass, (c) mild steel, (d) polyethylene, (e) rubber*

Table 2.1 *Tensile strength values at about 20°C*

Strength (MPa)	Material
2 to 12	Woods perpendicular to the grain
2 to 12	Elastomers
6 to 100	Woods parallel to the grain
60 to 100	Engineering polymers
20 to 60	Concrete
80 to 300	Magnesium alloys
160 to 400	Zinc alloys
100 to 600	Aluminium alloys
80 to 1000	Copper alloys
250 to 1300	Carbon and low alloy steels
250 to 1500	Nickel alloys
500 to 1800	High alloy steels
100 to 1800	Engineering composites
1000 to >10 000	Engineering ceramics

Example

A material has a yield stress of 200 MPa. What tensile forces will be needed to cause yielding with a bar of the material with a cross-sectional area of 100 mm^2 ?

$$\text{Yield force} = \text{yield stress} \times \text{area} = 200 \times 10^6 \times 100 \times 10^{-6} = 20\,000 \text{ N}$$

Example

Samples are taken of cast aluminium alloys and gave the following data: tensile strength LM4 140 MPa, LM6 160 MPa, LM9 170 MPa. Which is the strongest in tension?

The strongest in tension is LM9 with the highest tensile strength.

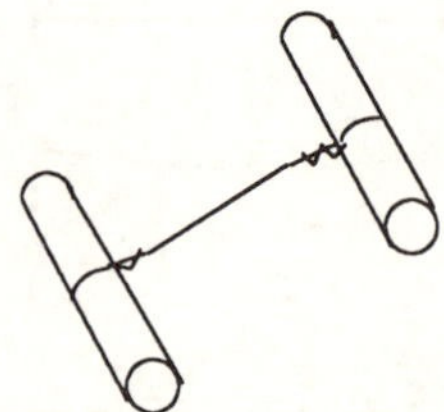

Figure 2.9 *Pulling wires*

Activity

A simple appreciation of the behaviour of a range of materials can be obtained by just pulling the materials between your hands and feeling how they behave. Materials that might be pulled in this way include a rubber band, polyethylene from a food bag, nylon from a fishing line, copper wire, steel wire, nichrome wire, fuse wire, etc. In the case of the wires, the ends should be twisted round a pair of dowel rods (Figure 2.9). *Safety note*: when doing experiments involving the stretching of wires, filaments, glass fibres or other materials, the specimen may fly up into your face when it breaks. *Safety spectacles should be worn.* Classify the materials in terms of strength, elastic behaviour and plastic behaviour.

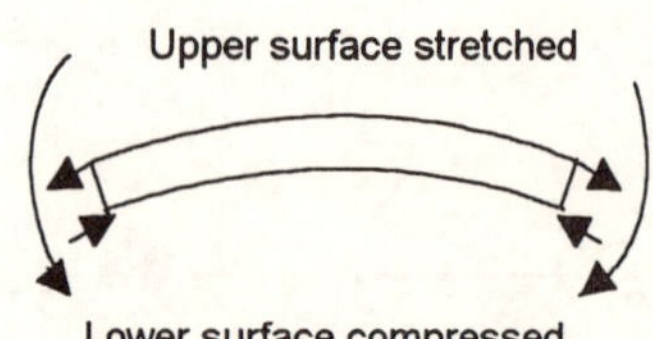

Figure 2.10 *Bending*

2.3.3 Stiffness

The *stiffness* of a material is the ability of a material to resist bending. When a strip of material is bent, one surface is stretched and the opposite face is compressed, as illustrated in Figure 2.10. The more a material

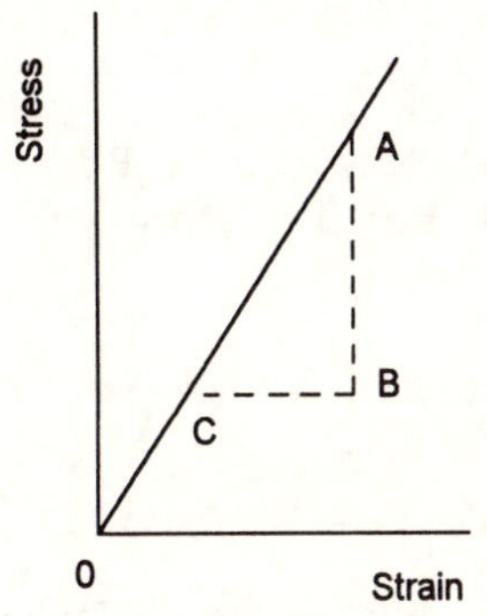

Figure 2.11 *Modulus of elasticity = AB/BC*

bends the greater is the amount by which the stretched surface extends and the compressed surface contracts. Thus a stiff material would be one that gave a small change in length when subject to tensile or compressive forces. This means a small strain when subject to tensile or compressive stress and so a small value of strain/stress, or conversely a large value of stress/strain. For most materials a graph of stress against strain gives initially a straight line relationship, as illustrated in Figure 2.11. Thus a large value of stress/strain means a steep slope of the stress-strain graph. The quantity stress/strain when we are concerned with the straight line part of the stress-strain graph is called the *modulus of elasticity* (or sometimes *Young's modulus*).

$$\text{Modulus of elasticity} = \frac{\text{stress}}{\text{strain}}$$

The units of the modulus are the same as those of stress, since strain has no units. Engineering materials frequently have a modulus of the order of 1000 000 000 Pa, i.e. 10^9 Pa. This is generally expressed as GPa, with 1 GPa = 10^9 Pa. Typical values are about 200 GPa for steels and about 70 GPa for aluminium alloys. The stress–strain graphs in Figure 2.7 give values of the initial slopes of the graphs and hence modulus of elasticity values. A stiff material has a high modulus of elasticity. Thus steels are stiffer than aluminium alloys. For most engineering materials the modulus of elasticity is the same in tension as in compression. Table 2.2 gives some typical values of tensile modulus at about 20°C.

Table 2.2 *Tensile modulus values at 20°C*

Tensile modulus (GPa)	Material
<0.2	Elastomers
0.2 to 10	Woods parallel to grain
0.2 to 10	Engineering polymers
2 to 20	Woods perpendicular to grain
20 to 50	Concrete
40 to 45	Magnesium alloys
50 to 80	Glasses
70 to 80	Aluminium alloys
43 to 96	Zinc alloys
110 to 125	Titanium alloys
100 to 160	Copper alloys
200 to 210	Steels
80 to 1000	Engineering ceramics

Example

For a material with a tensile modulus of elasticity of 200 GPa, what strain will be produced by a stress of 4 MPa?

Since the modulus of elasticity is stress/strain then:

$$\text{strain} = \frac{\text{stress}}{\text{modulus}} = \frac{4 \times 10^6}{200 \times 10^9} = 0.000\,02$$

Example

The tensile modulus of number of plastics is: ABS 2.5 GPa, PVC 3.1 GPa, polycarbonate 2.8 GPa, polypropylene 1.3 GPa. Which of the plastics is the stiffest?

The stiffest plastic is the one with the highest tensile modulus and so is the PVC.

Activity

Bend strips of wood, plastic and a metal, e.g. different forms of rulers, and list them in terms of tensile modulus.

2.3.4 Ductility/brittleness

Glass is a *brittle material* and if you drop a glass it breaks; however it is possible to stick all the pieces together again and restore the glass to its original shape. If a car is involved in a collision, the bodywork of mild steel is less likely to shatter like the glass but more likely to dent and show permanent deformation (the term *permanent deformation* is used for changes in dimensions which are not removed when the forces applied to the material are taken away). Materials which develop significant permanent deformation before they break are called *ductile*. Ductile materials permit manufacturing methods which involve bending them to the required shapes or using a press to squash the material into the required shape. Brittle materials cannot be formed to shape in this way.

The stress–strain graph for cast iron (Figure 2.8(a)) shows that very little plastic deformation occurs, no sooner has the stress risen to the yield point then failure occurs. Thus the length of a piece of cast iron after breaking is not much different from the initial length. Figure 2.8(c) shows the stress–strain graph for mild steel and this shows a considerable amount of plastic strain before breaking; it is ductile.

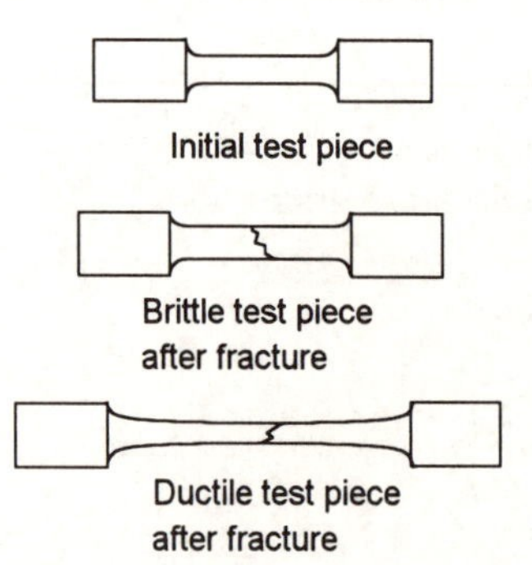

Figure 2.12 *Brittle and ductile test pieces after fracture*

A measure of the ductility of a material is obtained by determining the length of a test piece of the material, then stretching it until it breaks and then, by putting the pieces together, measuring the final length of the test piece (Figure 2.12). A brittle material will show little change in length from that of the original test piece, but a ductile material will indicate a significant increase in length. The *percentage elongation* of a test piece after breaking is thus used as a measure of ductility:

$$\%\ \text{elongation} = \frac{\text{final length} - \text{initial length}}{\text{initial length}} \times 100\%$$

A reasonably ductile material, such as mild steel, will have a percentage elongation of about 20%, a brittle material such as a cast iron less than 1%. Thermoplastics tend to have percentage elongations of the order of 50 to 500%, thermosets of the order of 0.1 to 1%. Thermosets are brittle materials, thermoplastics generally not.

Example
A 200 mm length of a material has a percentage elongation of 10%, by how much longer will a strip of the material be when it breaks?

$$\text{Change in length} = \frac{\%\ \text{elongation} \times \text{original length}}{100}$$

$$= \frac{10 \times 200}{100} = 20\ \text{mm}$$

Example
Which of the following brasses is the most ductile: 80–20 brass, percentage elongation 50%, 70–30 brass, percentage elongation 70%, 60–40 brass, percentage elongation 40%?

The most ductile material is the one with the largest percentage elongation, i.e. the 70–30 brass.

Example
A sample of a carbon steel has a tensile strength of 400 MPa and a percentage elongation of 35%. A sample of an aluminium–manganese alloy has a tensile strength of 140 MPa and a percentage elongation of 10%. What does this data tell you about the mechanical behaviour of the materials?

The higher value of the tensile strength of the carbon steel indicates that the material is stronger and, for the same cross-sectional area, a bar of carbon steel could withstand higher tensile forces than a corresponding bar of the aluminium alloy. The higher percentage elongation of the carbon steel indicates that the material has a greater ductility than the aluminium alloy. Indeed the value is such as to indicate that the carbon steel is very ductile. The steel is thus stronger and more ductile.

2.3.5 Toughness

The materials in many products may contain cracks or sharp corners or other changes in shape that can readily generate cracks. A tough material can be considered to be one that, though it may contain a crack, resists breaking as a result of the crack growing and running through the material. Think of trying to tear a sheet of paper. If there is an initial 'crack' then the material is much more easily torn. In the case of the paper, the initial 'cracks' may be perforations put there to enable the paper to be torn easily. In the case of, say, the skin of an aircraft where there may be holes, such as windows or their fastenings, which are equivalent to cracks, there is a need for cracks not to propagate. A tough material is required.

Toughness can be defined in terms of the work that has to be done to propagate a crack through a material, a tough material requiring more energy than a less tough one. Consider a length of material being stretched by tensile forces. When a length of material is stretched by an amount x_1 as a result of a constant force F_1 then the work done is the force × distance moved by point of application of force and thus:

$$\text{work} = F_1 x_1$$

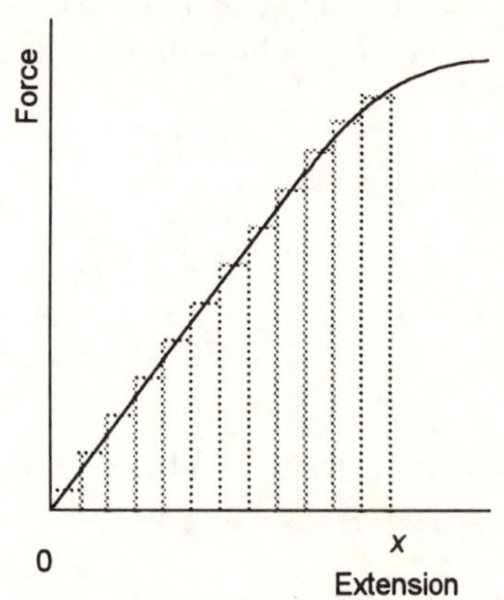

Figure 2.13 *Area under force–extension graph*

Thus if a force–extension graph is considered (Figure 2.13), the work done, when we consider a very small extension, is the area of that strip under the graph. The total work done in stretching a material to an extension x, i.e. through an extension which we can consider to be made up of a number of small extensions with $x = x_1 + x_2 + x_3 + \ldots$, is thus:

$$\text{work} = F_1 x_1 + F_2 x_2 + F_3 x_3 + \ldots$$

and so is the area under the graph up to x. If we divide both sides of this equation by the volume, i.e. the product of the cross-sectional area A of the strip and its length L, we have:

$$\frac{\text{work}}{\text{volume}} = \left(\frac{F_1}{A} \times \frac{x_1}{L}\right) + \left(\frac{F_2}{A} \times \frac{x_2}{L}\right) + \left(\frac{F_3}{A} \times \frac{x_3}{L}\right) + \ldots$$

But the term in each bracket is just the product of the stress and strain. Thus the work done per unit volume of material is the area under the stress–strain graph up to the strain corresponding to extension x. The area under the stress–strain graph up to some strain is the energy required per unit volume of material to produce that strain. For a crack to propagate, a material must fail. Thus the area under the stress–strain graph up to the breaking point is a measure of the energy required to break unit volume of the material and so for a crack to propagate. A large area is given by a material with a large yield stress and high ductility. Such materials can thus be considered to be tough.

An alternative way of considering toughness is the ability of a material to withstand shock loads. A measure of this ability to withstand suddenly applied forces is obtained by *impact tests*, such as the Charpy and Izod tests (see Chapter 4). In these tests, a test piece is struck a sudden blow and the energy needed to break it is measured. A brittle material will require less energy than a ductile material. The results of such tests are often used as a measure of the brittleness of materials.

Another measure of toughness that can be used is fracture toughness. *Fracture toughness* can be defined as a measure of the ability of a material to resist the propagation of a crack. The toughness is determined by loading a sample of the material which contains a deliberately introduced crack of length $2c$ and recording the tensile stress σ at which the crack propagates. The fracture toughness, symbol K_c and usual units MPa $m^{1/2}$, is given by

$$K_c = \sigma\sqrt{\pi c}$$

The smaller the value of the toughness the more readily a crack propagates. The value of the toughness depends on the thickness of the material, high values occurring for thin sheets and decreasing with increasing thickness to become almost constant with thick sheets. For this reason a value called the *plane strain fracture toughness* K_{Ic} is often quoted. This is the value of the toughness that would be obtained with thick sheets. Typical values are of the order of 1 MPa $m^{1/2}$ for glass, which readily fractures when there is a crack present, to values of the order of 50 to 150 MPa $m^{1/2}$ for some steels and copper alloys. In such materials cracks do not readily propagate.

2.3.6 Hardness

The *hardness of a material* is a measure of the resistance of a material to abrasion or indentation. A number of scales are used for hardness, depending on the method that has been used to measure it (see Chapter 4 for discussions of test methods). The hardness is roughly related to the tensile strength of a material, the tensile strength being roughly proportional to the hardness (see Chapter 4). Thus the higher the hardness of a material the higher is likely to be the tensile strength.

2.4 Electrical properties

The electrical *resistivity* ρ is a measure of the electrical resistance of a material, being defined by the equation

$$\rho = \frac{RA}{L}$$

where R is the resistance of a length L of the material of cross-sectional area A (Figure 2.14). The unit of resistivity is the ohm metre. An electrical *insulator*, such as a ceramic, will have a very high resistivity, typically of the order of 10^{10} Ω m or higher. An electrical *conductor*, such as copper, will have a very low resistivity, typically of the order of 10^{-8} Ω m. The term *semiconductor* is used for materials with resistivities roughly half-way between those of conductors and insulators, i.e. about 10^2 Ω m.

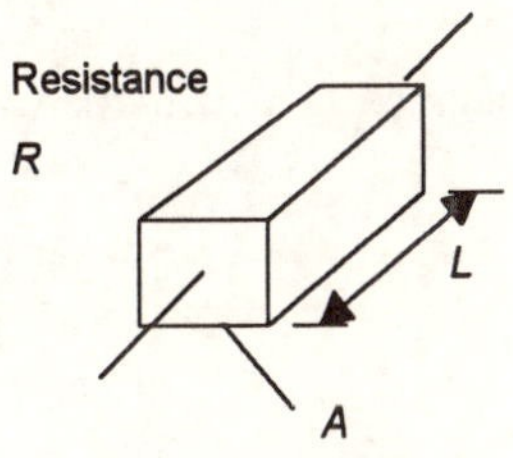

Figure 2.14 *Resistivity*

The electrical *conductance* of a length of material is the reciprocal of its resistance and has the unit of Ω^{-1}. This unit is given a special name, the siemen (S). The electrical *conductivity* σ is the reciprocal of the resistivity, i.e.

$$\sigma = \frac{1}{\rho} = \frac{L}{RA}$$

The unit of conductivity is thus Ω^{-1} m^{-1} or S m^{-1}. Since conductivity is the reciprocal of the resistivity, an electrical insulator will have a very low conductivity, of the order of 10^{-10} S/m, while an electrical conductor will have a very high conductivity, of the order of 10^8 S/m.

Table 2.3 shows typical values of resistivity and conductivity for insulators, semiconductors and conductors. Pure metals and many metal alloys have resistivities that increase when the temperature increases; some metal alloys do, however, show increases in resistivities when the temperature increases. For semiconductors and insulators, the resistivity increases with an increase in temperature.

Table 2.3 *Typical resistivity and conductivity values at about 20°C*

Material	Resistivity Ω m	Conductivity S/m
Insulators		
Acrylic (a polymer)	$> 10^{14}$	$< 10^{-14}$
Polyvinyl chloride (a polymer)	$10^{12}-10^{13}$	$10^{-13}-10^{-12}$
Mica	$10^{11}-10^{12}$	$10^{-12}-10^{-11}$
Glass	$10^{10}-10^{14}$	$10^{-14}-10^{-10}$
Porcelain (a ceramic)	$10^{10}-10^{12}$	$10^{-12}-10^{-10}$
Alumina (a ceramic)	$10^{9}-10^{12}$	$10^{-12}-10^{-9}$
Semiconductors		
Silicon (pure)	2.3×10^{3}	4.3×10^{-4}
Germanium (pure)	0.43	2.3
Conductors		
Nichrome (alloy of nickel and chromium)	108×10^{-8}	0.9×10^{6}
Manganin (alloy of copper and manganese)	42×10^{-8}	2×10^{6}
Nickel (pure)	7×10^{-8}	14×10^{6}
Copper (pure)	2×10^{-8}	50×10^{6}

Example
Using the value of electrical conductivity given in Table 2.3, determine the electrical conductance of a 2 m length of nichrome wire at 20°C if it has a cross-sectional area of 1 mm².

Using the equation $\sigma = L/RA$, with the conductance $G = 1/R$, then we have $\sigma = LG/A$ and so:

$$G = \frac{\sigma A}{L} = \frac{0.9 \times 10^{6} \times 1 \times 10^{-6}}{2} = 0.45 \text{ S}$$

Example
Suggest a material that could be used for the heating element of an electric fire?

The heating element must be a conductor of electricity. The power dissipated by the element is V^2/R, thus the lower the resistance R the greater the power produced by a given voltage V. The material must also be able to withstand high temperatures without melting or oxidising. Nichrome wire is commonly used. The wire is wound on a spiral around an insulating ceramic support.

Activity
Dismantle an electric plug and identify the types of materials used and whether they are insulators or conductors.

2.4.1 Dielectric strength

The *dielectric strength* is a measure of the highest voltage that an insulating material can withstand without electrical breakdown. It is defined as:

$$\text{dielectric strength} = \frac{\text{breakdown voltage}}{\text{insulator thickness}}$$

The units of dielectric strength are volts per metre. Polythene has a dielectric strength of about 4×10^7 V/m. This means that a 1 mm thickness of polythene will require a voltage of about 40 000 V across it before it will break down.

Example
An electrical capacitor is to be made with a sheet of polythene of thickness 0.1 mm between the capacitor plates. What is the greatest voltage that can be connected between the capacitor plates if there is not to be electrical breakdown and the dielectric strength is 4×10^7 V/m?

The dielectric strength is defined as the breakdown voltage divided by the insulator thickness, hence

$$\text{breakdown voltage} = \text{dielectric strength} \times \text{thickness}$$

$$= 4 \times 10^7 \times 0.1 \times 10^{-3} = 4000 \text{ V}$$

2.5 Thermal properties

Thermal properties that are generally of interest in the selection of materials include how much a material will expand for a particular change in temperature; how much the temperature of a piece of material will change when there is a heat input into it; and how good a conductor of heat it is. The SI unit of temperature is the kelvin (K), with a temperature change of 1 K being the same as a change of 1°C.

The *linear expansivity* α or *coefficient of linear expansion* is a measure of the amount by which a length of material expands when the temperature increases. It is defined as:

$$\alpha = \frac{\text{change in length}}{\text{original length} \times \text{change in temperature}}$$

and has the unit of K^{-1}.

The term *heat capacity* is used for the amount of heat needed to raise the temperature of an object by 1 K. Thus if 300 J is needed to raise the temperature of a block of material by 1 K, then its heat capacity is 300 J/K. The *specific heat capacity* c is the amount of heat needed per kilogram of material to raise the temperature by 1 K, hence:

$$c = \frac{\text{amount of heat}}{\text{mass} \times \text{change in temperature}}$$

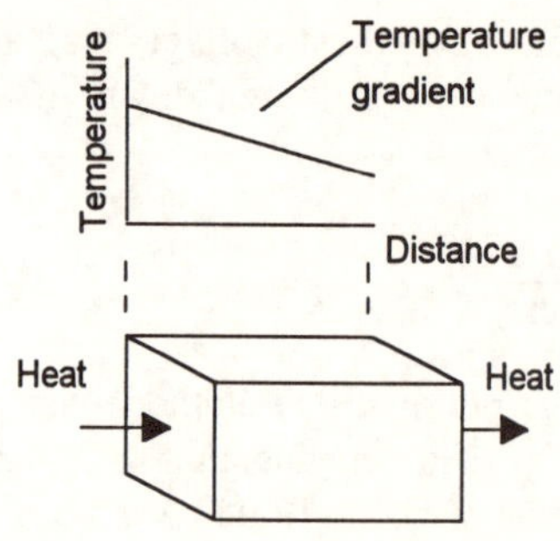

Figure 2.15 *Thermal conductivity*

It has the unit of J kg^{-1} K^{-1}. Weight-for-weight, metals require less heat to reach a particular temperature than plastics, e.g. copper has a specific heat capacity of about 340 J kg^{-1} K^{-1} while polythene is about 1800 J kg^{-1} K^{-1}.

The *thermal conductivity* λ of a material is a measure of the ability of a material to conduct heat. There will only be a net flow of heat energy through a length of material when there is a difference in temperature between the ends of the material. Thus the thermal conductivity is defined in terms of the quantity of heat that will flow per second divided by the temperature gradient (Figure 2.15), i.e.:

$$\lambda = \frac{\text{quantity of heat/second}}{\text{temperature gradient}}$$

and has the unit of W m^{-1} K^{-1}. A high thermal conductivity means a good conductor of heat. Metals tend to be good conductors, e.g. copper has a thermal conductivity of about 400 W m^{-1} K^{-1}. Materials that are bad conductors of heat have low thermal conductivities, e.g. plastics have thermal conductivities of the order 0.3 W m^{-1} K^{-1} or less. Very low thermal conductivities occur with foamed plastics, i.e. those containing bubbles of air. Foamed polymer polystyrene (expanded polystyrene) has a thermal conductivity of about 0.02 to 0.03 W m^{-1} K^{-1} and is widely used for thermal insulation.

Table 2.4 gives typical values of the linear expansivity, the specific heat capacity and the thermal conductivity for metals, polymers and ceramics.

Activity

By just touching a number of surfaces, list them in terms of their thermal conductivy, explaining your reasons for doing so.

Table 2.4 *Thermal properties*

Material	Linear expansivity 10^{-6} K^{-1}	Specific heat capacity J kg^{-1} K^{-1}	Thermal conductivity W m^{-1} K^{-1}
Metals			
Aluminium	24	920	230
Copper	18	385	380
Mild steel	11	480	54
Polymers			
Polyvinyl chloride	70–80	840–1200	0.1–0.2
Polyethylene	100–200	1900–2300	0.3–0.5
Epoxy cast resin	45–65	1000	0.1–0.2
Ceramics			
Alumina	8	750	38
Fused silica	0.5	800	2
Glass	8	800	1

Example

By how much will a 10 cm strip of (a) copper, (b) PVC expand when the temperature changes from 20 to 30°C? Use the data given in Table 2.4.

(a) For copper:

$$\text{expansion} = 18 \times 10^{-6} \times 0.10 \times 10 = 19 \times 10^{-6}\ \text{m} = 0.018\ \text{mm}$$

(b) For PVC:

$$\text{expansion} = 75 \times 10^{-6} \times 0.10 \times 10 = 75 \times 10^{-6}\ \text{m} = 0.075\ \text{mm}$$

The expansion for PVC is some four times greater than that for copper.

Example

The heating element for an electric fire is wound on an electrical insulator. What thermal considerations will affect the choice of insulator material?

The insulator will need to have a low heat capacity so that little heat is used to raise the material to temperature. This means using a material with as low a density, and hence low mass, and specific heat capacity as possible. It also will need to be able to withstand the temperatures realised without deformation or melting. A ceramic is indicated.

2.6 Physical properties

The *density* ρ of a material is the mass per unit volume:

$$\rho = \frac{\text{mass}}{\text{volume}}$$

It has the unit of kg/m^3. It is often an important property that is required in addition to a mechanical property. Thus, for example, an aircraft undercarriage is required to be not only strong but of low mass. Thus what is required is as high a strength as possible with as low a density as possible. Thus what is looked for is a high value of strength/density. This quantity is often referred to as the *specific strength*. Table 2.5 gives some typical values. A magnesium alloy might thus be preferred to a steel for the undercarriage, such an alloy often having a higher specific strength.

2.7 Durability properties

Attack of materials by the environment in which they are situated is a major problem. The rusting of iron is an obvious example. Tables, e.g. Table 2.6, are often used giving the comparative resistance to attack of materials in various environments, e.g. in aerated water, in salt water, to strong acids, to strong alkalis, to organic solvents, to ultraviolet radiation. Thus, for example, in a salt water environment carbon steels are rated at having very poor resistance to attack, aluminium alloys good resistance and stainless steels excellent resistance.

Table 2.5 *Typical strength/density values at about 20°C*

Material	Density (Mg/m^3)	Strength to density ratio ($MPa/Mg\ m^{-3}$)
Aluminium alloys	2.6 to 2.9	40 to 220
Copper alloys	7.5 to 9.0	8 to 110
Lead alloys	8.9 to 11.3	1 to 3
Magnesium alloys	1.9	40 to 160
Nickel alloys	7.8 to 9.2	30 to 170
Titanium alloys	4.3 to 5.1	40 to 260
Zinc alloys	5.2 to 7.2	30 to 60
Carbon and low alloy steels	7.8	30 to 170
High alloy steels	7.8 to 8.1	60 to 220
Engineering ceramics	2.2 to 3.9	>300
Glasses	2 to 3	200 to 800
Thermoplastics	0.9 to 1.6	15 to 70
Polymer foams	0.04 to 0.7	0.4 to 12
Engineering composites	1.4 to 2	70 to 900
Concrete	2.4 to 2.5	8 to 30
Wood	0.4 to 1.8	5 to 60

Table 2.6 *Compative corrosion resistances*

Corrosion resistance	Material
Aerated water	
High resistance	All ceramics, glasses, lead alloys, alloy steels, titanium alloys, nickel alloys, copper alloys, PTFE, polypropylene, nylon, epoxies, polystyrene, PVC
Medium resistance	Aluminium alloys, polythene, polyesters
Low resistance	Carbon steels
Salt water	
High resistance	All ceramics, glasses, lead alloys, stainless steels, titanium alloys, nickel alloys, copper alloys, PTFE, polypropylene, nylon, epoxies, polystyrene, PVC, polythene
Medium resistance	Aluminium alloys, polyesters
Low resistance	Low alloy steels, carbon steels
UV radiation	
High resistance	All ceramics, glasses, all alloys
Medium resistance	Epoxies, polyesters, polypropylene, polystyrene, HD polyethylene, polymers with UV inhibitor
Low resistance	Nylon, PVC, many elastomers

2.8 The range of materials

Table 2.7 lists the range of properties that are typical of metals, polymers and ceramics (1 Mg/m^3 = 1000 kg/m^3).

Table 2.7 *The range of properties*

Property	Metals	Polymers	Ceramics
Density Mg/m^3	2–16	1–2	2–17
Melting point °C	200–3500	70–200	2000–4000
Thermal conductivity	High	Low	Medium
Thermal expansion	Medium	High	Low
Specific heat capacity	Low	Medium	High
Electrical conductivity	High	Very low	Very low
Tensile strength MPa	100–2500	30–300	10–400
Tensile modulus GPa	40–400	0.7–3.5	150–450
Hardness	Medium	Low	High
Resistance to corrosion	Medium–poor	Good–medium	Good

2.9 Costs

Costs can be considered in relation to the basic costs of the raw materials, the costs of manufacturing, and the life and maintenance costs of the finished product.

Comparison of the basic costs of materials is often on the basis of the cost per unit weight or cost per unit volume. If the cost of 10 kg of a metal is, say, £1 then the cost per kg is £0.1. If the metal has a density of 8000 kg/m^3 then 10 kg will have a volume of 10/8000 = 0.001 25 m^3 and so the cost per cubic metre is 1/0.001 25 = £800. Thus we can write

$$\text{cost per m}^3 = (\text{cost/kg}) \times \text{density}$$

Table 2.8 shows the relative costs per kg and per cubic metre of some materials.

However, often a more important comparison is on the basis of the *cost per unit strength* or *cost per unit stiffness* for the same volume of material. This enables the cost of say a beam to be considered in terms of what it will cost to have a beam of a certain strength or stiffness.

Hence if, for comparison purposes, we consider a beam of volume 1 m^3 then if the tensile strength of the above material is 500 MPa the cost per MPa of strength will be 800/500 = £1.6. Thus we can write, for the same volume:

$$\text{cost/unit strength} = \frac{(\text{cost/m}^3)}{\text{strength}}$$

and similarly:

$$\text{cost/unit stiffness} = \frac{(\text{cost/m}^3)}{\text{modulus}}$$

Table 2.8 *Relative costs of materials in relation to that of mild steel*

Material	Relative cost/kg	Relative cost/m³
Nickel	28	32
Chromium	26	24
Tin	19	18
Brass sheet	16	17
Al–Cu alloy sheet	14	5.3
Nylon 66	12	1.8
Magnesium ingot	9.2	2.1
Acrylic	8.9	1.4
Copper tubing	8.7	10
ABS	8.3	1.1
Aluminium ingot	4.3	1.5
Polystyrene	3.6	0.50
Zinc ingot	3.6	3.3
Polyethylene (HDPE)	3.4	0.43
Polypropylene	3.2	0.34
Natural rubber	3.1	0.50
Polyethylene (LDPE)	2.3	0.29
PVC, rigid	2.3	0.43
Mild steel sheet	1.9	1.9
Mild steel ingot	1.0	1.0
Cast iron	0.8	0.79

The costs of manufacturing will depend on the processes used. Some processes require a large capital outlay and then can be used to produce large numbers of the product at a relatively low cost per item. Other processes may have little in the way of setting-up costs but a large cost per unit product.

The cost of maintaining a material during its life can often be a significant factor in the selection of materials. A feature common to many metals is the need for a surface coating to protect them for corrosion by the atmosphere. The rusting of steels is an obvious example of this and dictates the need for such activities as the continuous repainting of the Forth Railway Bridge.

Example

On the basis of the following data, compare the costs per unit strength of the two materials for the same volume of material.

Low carbon steel:
Cost per kg £0.1, density 7800 kg/m^3, strength 1000 MPa
Aluminium alloy (Mn):
Cost per kg £0.22, density 2700 kg/m^3, strength 200 MPa

For the steel, the volume of 1 kg is 1/7800 = 0.000 13 m^3 and so the cost per m^3 is 0.1/0.00013 = £770. The cost per MPa of strength is thus 770/1000 = £0.77. For the aluminium alloy, the volume of 1 kg is 1/2700

= 0.000 37 m^3 and so the cost per m^3 is 0.22/0.000 37 = £590. Thus although the cost per kg is greater than that of the steel, because of the lower density the cost per cubic metre is less. The cost per MPa of strength is 590/200 = £2.95. Hence on a comparison on the strengths of equal volumes, it is cheaper to use the steel.

Activity

Write notes about the properties that might be significant in the use of mild steel for the bodywork of cars.

Problems

1 What types of properties would be required for the following products?
   (a) A domestic kitchen sink.
   (b) A shelf on a bookcase.
   (c) A cup.
   (d) An electrical cable.
   (e) A coin.
   (f) A car axle.
   (g) The casing of a telephone.
2 For each of the products listed in problem 1, identify a material that is commonly used and explain why its properties justify its choice for that purpose.
3 Which properties of a material would you need to consider if you required materials which were:
   (a) stiff,
   (b) capable of being bent into a fixed shape,
   (c) capable of not fracturing when small cracks are present,
   (d) not easily breaking,
   (e) acting as an electrical insulator,
   (f) a good conductor of heat,
   (g) capable of being used as the lining for a tank storing acid.
4 A colleague informs you that a material has a high tensile strength with a low percentage elongation. Explain how you would expect the material to behave.
5 A colleague informs you that a material has a high tensile modulus of elasticity and good fracture toughness. Explain how you would expect the material to behave.
6 What is the tensile stress acting on a strip of material of cross-sectional area 50 mm^2 when subject to tensile forces of 1 kN?
7 Tensile forces act on a rod of length 300 mm and cause it to extend by 2 mm. What is the strain?
8 An aluminium alloy has a tensile strength of 200 MPa. What force is needed to break a bar of this material with a cross-sectional area of 250 mm^2?
9 A test piece of a material is measured as having a length of 100 mm before any forces are applied to it. After being subject to tensile forces it breaks and the broken pieces are found to have a combined length of 112 mm. What is the percentage elongation?

10 A material has a yield stress of 250 MPa. What tensile forces will be needed to cause yielding if the material has a cross-sectional area of 200 mm^2?

11 A sample of high tensile brass is quoted as having a tensile strength of 480 MPa and a percentage elongation of 20%. An aluminium-bronze is quoted as having a tensile strength of 600 MPa and a percentage elongation of 25%. Explain the significance of this data in relation to the mechanical behaviour of the materials.

12 A grey cast iron is quoted as having a tensile strength of 150 MPa, a compressive strength of 600 MPa and a percentage elongation of 0.6%. Explain the significance of the data in relation to the mechanical behaviour of the material.

13 A sample of a carbon steel is found to have an impact energy of 120 J at temperatures above 0°C and 5 J below it. What is the significance of this data?

14 Mild steel is quoted as having an electrical resistivity of 1.6×10^{-7} Ω m. Is it a good conductor of electricity?

15 A colleague states that he needs a material with a high electrical conductivity. Electrical resistivity tables for materials are available. What types of resistivity values would you suggest he looks for?

16 Aluminium has a resistivity of 2.5×10^{-8} Ω m. What will be the resistance of an aluminium wire with a length of 1 m and a cross-sectional area of 2 mm^2?

17 How do the properties of thermoplastics differ from those of thermosets?

18 You read in a textbook that 'Designing with ceramics presents problems that do not occur with metals because of the almost complete absence of ductility with ceramics'. Explain the significance of the comment in relation to the exposure of ceramics to forces.

19 Compare the specific strengths, and costs per unit strength for equal volumes, for the materials giving the following data:

Low carbon steel:
Cost per kg £0.1, density 7800 kg/m^3, strength 1000 MPa
Polypropylene:
Cost per kg £0.2, density 900 kg/m^3, strength 30 MPa

3 Properties data

3.1 Standards

There are many thousands of standards laid down by national standards bodies such the British Standards Association (BS) and international bodies such as the International Organisation for Standardisation (ISO). A *standard* is a technical specification drawn up with the co-operation and general approval of all interests affected by it with the aims of benefiting all concerned by the promotion of consistency in quality, rationalisation of processes and methods of operation, promoting economic production methods, providing a means of communication, protecting consumer interests, promoting safe practices, and helping confidence in manufacturers and users. Thus, for example, there is a standard on the design of cylindrical and spherical pressure vessels subject to internal pressure (BS 5500). This standard covers the construction of pressure vessels for use in chemical plants and gives equations and software solutions by which the wall thickness can be calculated in order to ensure safe operation. There is a standard for the graphical symbols used in technical drawing (BS 1553). This specifies the form symbols used in drawing should take so that all those reading the drawings will understand their significance.

Standards, both national and international, are used in the case of materials to ensure such things as consistency of quality, consistency in the use of terms, rationalisation of testing methods, and provide an efficient means of communication between interested parties. Thus if a material is stated by its producer to be to a certain standard, tested by the methods laid down by certain standards, then a customer need not have all the details written out of what properties and tests have been carried out by the producer in order to know what properties to expect of the material. There is, for example, the standard for tensile testing of metals (BSEN 10002, a European standard adopted as the British Standard and replacing BS 18). This lays down such things as the sizes of the test pieces to be used (see Chapter 3 for such details). There are standards for materials such as copper and copper alloy plate (BS 2875), steel plate, sheet and strip (BS 1449), the plastic polypropylene (BS 5139), etc. which lay down the composition and properties for such materials.

3.2 Data sources

Data on the properties of materials is available from a range of sources. These include:

1 Specifications issued by bodies responsible for standards, e.g. the British Standards Institution, the American Society for Metals, the

International Organisation for Standardisation, etc. The standards operating in Britain are those issued by the British Standards Institution. Their catalogue lists all the British Standards and also whether they are also European and/or international standards.

2 Data books, e.g. the *ASM Metals Reference Book* (American Society for Metals 1983), *Metals Reference Book* by R.J. Smithells (Butterworth 1987), *Metals Databook* by C. Robb (The Institute of Metals 1987), *Handbook of Plastics and Elastomers* edited by C.A. Harper (McGraw-Hill 1975), *Newnes Engineering Materials Pocket Book* by W. Bolton (Heinemann-Newnes 1989, 1996).

3 Computerised data bases which give materials and their properties with means to rapidly access particular materials or find materials with particular properties, e.g. Cambridge Materials Selector (Cambridge University Engineering Department 1992), MAT.DB (American Society for Metals 1990).

4 Trade associations, e.g. such bodies as the Copper Development Association which issues brochures giving technical details of compositions and properties of copper and copper alloy materials.

5 Data sheets supplied by suppliers of materials.

6 In-company tests. This is more often than not used to check samples of a bought-in material to ensure that it conforms to the standards specified by the supplier from which the material was bought.

3.3 Coding systems

Coding systems are used to refer to particular metals. Such codes can relate to the chemical composition or the type of properties it has and are a concise way of specifying a particular material without having to write out its full chemical composition or properties.

3.3.1 Steels

Steels might be referred to in terms of a code specified by the British Standards Association. The code for wrought steels is a six symbol code:

The first three digits of the code represent the type of steel

000 to 199	Carbon and carbon-manganese types, the number being 100 times the manganese contents
200 to 240	Free-cutting steels, the second and third numbers being approximately 100 times the mean sulphur content
250	Silicon-manganese spring steels
300 to 499	Stainless and heat resistant valve steels
500 to 999	Alloy steels with different groups of numbers within this range being allocated to different alloys. Thus 500 to 519 is when nickel is the main alloying element, 520 to 539 when chromium is.

The fourth symbol is a letter

A The steel is supplied to a chemical composition determined by chemical analysis

H The steel is supplied to a hardenability specification

M The steel is supplied to mechanical property specifications

S The steel is stainless

The fifth and sixth digits correspond to 100 times the mean percentage carbon content of the steel.

As an illustration of the above coding system, consider a steel with a code 070M20. The first three digits are 070 and since they are between 000 and 199 the steel is a carbon or carbon-manganese type. The 070 indicates that the steel has 0.70% manganese. The fourth symbol is M and so the steel is supplied to mechanical property specifications. The fifth and sixth digits are 20 and so the steel has 0.20% carbon.

The AISI-SAE (American Iron and Steel Institute, Society of Automotive Engineers) use a four digit code. The first two digits indicate the type of steel. The third and fourth digits are used to indicate 100 times the percentage of the carbon content.

1000	Carbon steel
	Manganese steels 1300
2000	Nickel steels
3000	Nickel–chromium steels
4000	Molybdenum steels 4000, 4400
	Chromium–molybdenum steels 4100
	Nickel–chromium–molybdenum steels 4300, 4700
	Nickel–molybdenum steels 4600, 4800
5000	Chromium steels
6000	Chromium–vanadium steels
7000	Tungsten–chromium steels
8000	Nickel–chromium–molybdenum steels
9000	Silicon–manganese steels 9200
	Nickel–chromium–molybdenum steels 9300, 9400

For example, 1010 is a carbon steel with about 0.10% carbon; 5120 is a chromium steel with 0.20% carbon.

3.3.2 Aluminium alloys

Aluminium cast alloys use the code of LM followed by a number, the number being used to indicate a specific alloy listed in BS 1490. The coding system for wrought aluminium alloys is that of the Aluminium Association. This uses four digits with first digit representing the principal alloying element, the second modifications to impurity limits, and the last two digits for the 1XXX alloys the aluminium content above 99.00% in hundredths and for others as identification of specific alloys. The 1XXX alloys have the principal alloying element of 99.00%

minimum aluminium, the 2XXX copper as the principal alloying element, 3XXX manganese, 4XXX silicon, 5XXX magnesium, 6XXX magnesium and silicon, 7XXX zinc, 8XXX other elements.

3.3.3 Copper alloys

A commonly used system is that of the Copper Development Association (CDA). This uses the letter C followed by three digits. The first digit indicates the group of alloys concerned and the remaining two digits alloys within the group. The groups C1XX to C7XX are used for wrought alloys and C8XX and C9XX for cast alloys.

C1XX	Copper with a minimum copper content of 99.3%. High copper alloys, with more than 96% copper
C2XX	Copper–zinc alloys (brasses)
C3XX	Copper–zinc–lead alloys (leaded brasses)
C4XX	Copper–zinc–tin alloys (tin brasses)
C5XX	Copper–tin alloys (phosphor bronzes)
C6XX	Copper–aluminium alloys (aluminium bronzes) Copper–silicon alloys (silicon bronzes) Miscellaneous copper–zinc alloys
C7XX	Copper–nickel Copper–nickel–zinc alloys (nickel silvers)
C8XX	Cast coppers, high copper alloys, brasses, manganese bronze and copper–zinc–silicon alloys
C9XX	Cast copper-tin, copper–tin–lead, copper–tin–nickel, copper–aluminium–iron, copper–nickel–iron and copper–nickel–zinc alloys

The British Standards Association coding for wrought copper and copper alloys consists of two letters followed by three digits. The two letters indicate the alloy group and the three digits the alloy within that group. Thus for copper and alloys containing a very high percentage of copper the letter used is C, for copper-aluminium alloys, i.e. aluminium bronzes, the letters are CA, for copper-beryllium alloys, i.e. beryllium bronzes, the letters are CB, for copper-nickel alloys, i.e. cupro-nickels, the letters are CN, for copper-silicon alloys, i.e. silicon bronzes, the letters are CS, for copper-zinc alloys, i.e. brasses, the letters are CZ, for copper-zinc-nickel alloys, i.e. nickel-silvers, the letters are NS, for copper-tin-phosphorus alloys, i.e. phosphor bronzes, the letters are PB. For cast copper alloys, letters followed by a digit are used. For example, AB1, AB2, and AB3 are aluminium bronze, LB1, LB2, LB4, and LB5 are leaded bronzes, SCB1 and SCB2 are general purpose brasses.

3.3.4 Temper

The above gives an indication of some of the codes developed by Standards bodies to assist communication between suppliers and users of materials. In addition, since the properties of materials depend on the heat treatment and degree of working they have undergone, there are codes to specify such treatments. The conditions being generally referred to as the *temper* of the material. Thus, for example, copper alloys are designated as temper O if the material is supplied in the annealed condition, H if supplied in the hard condition resulting from cold working with ¼H and ½H to indicate quarter and half-hard conditions, M for as manufactured, W(H) for the heat treatment called solution treatment to the hardened condition. With aluminium alloys M indicates as manufactured, O annealed, H followed by a number between 1 and 8 the degree of hardening resulting from cold working, T followed by various letters different forms of heat treatment.

3.4 Data analysis

The procedure that might be adopted in searching for a material with the properties required for a particular product might be:

1. Identify the properties required.
2. Look in British or International Standards, or data books or computer data bases for materials with the required properties.
3. The above might refine the search to a particular material. Then, if further information is required, trade association publications or supplier data sheets can be consulted.

To illustrate the above, consider the search for a material for use as a conductor of electricity where high conductivity is the main property required. Since metals are, in general good conductors and polymers and ceramics very poor conductors then the choice would seem to be among metals. When tables are consulted the following information can be found for electrical conductivities, at 20°C:

Aluminium	40×10^6 S/m
Copper	64×10^6 S/m
Gold	50×10^6 S/m
Iron	11×10^6 S/m
Silver	67×10^6 S/m

If costs is also a factor then gold and silver are likely to be ruled out. Thus copper looks to be the optimum choice. If we now consider what form of copper then tables are likely to yield data in the following form:

C101	Electrolytic tough-pitch h.c. copper	101.5-100
C103	Oxygen-free h.c. copper	101.5-100
C105	Phosphorus deoxidised arsenical copper	50-35
C108	Copper-cadmium	92-80

The conductivities are not expressed in S/m but in units called IACS (International Annealed Copper Standard) units and written as a percentage. This scale is based on 100% being the conductivity (or resistivity) of annealed copper at 20°C. This is a resistivity of $1.7241 \times 10^{-8}\ \Omega$ m or a conductivity of 58.00×10^6 S/m. Thus, if there are no other considerations C101 or C103 would appear to be the choice. Often, however, there are other factors, e.g. strength, to be taken into account.

Consider another example, a requirement for a metal with a low melting point. The aim is to use the metals in die casting for the production of small components for toys, e.g. toy car steering wheels and drive shafts. The following are some of the melting points for metals that can be found from tables:

Aluminium	600°C
Lead	320°C
Magnesium	520°C
Zinc	380°C

Lead and zinc have the lowest melting points. If we add the requirement of reasonable strength in the as-cast condition then tables give:

Lead	Tensile strength 20 MPa
Zinc	Tensile strength 280 MPa

Thus, taking strength into account, zinc would appear to be the choice. Other factors, such as the poisonous nature of lead, would also suggest that zinc is the choice.

Example

The data in Table 3.1 gives the properties of a number of cast irons. Select, from those listed, a cast iron which combines high tensile strength with ductility.

Table 3.1 *Mechanical properties of cast irons*

Material	Tensile strength MPa	Yield stress MPa	Percentage elongation
Grey irons			
BS 150	150	98	0.6
BS 180	180	117	0.5
BS 220	220	143	0.5
BS 260	260	170	0.4
BS 300	300	195	0.3
BS 350	350	228	0.3
BS 400	400	260	0.2
Malleable irons			
Blackheart B32-10	320	190	10
Blackheart B35-12	350	200	12
Whiteheart W38-12	380	200	12
Whiteheart W40-05	400	220	5
Whiteheart W45-07	450	260	7

A high ductility means a high percentage elongation. Grey irons have very low percentage elongations and so are brittle. The malleable irons shows greater percentage elongations and thus the selection needs to be one of these irons. The malleable iron with the greatest tensile strength Whiteheart W45-07 does not however have so high a percentage elongation as Whiteheart W38-12. Thus if ductility is more important than strength the choice might be Whiteheart W38-12.

Activity

Prepare a report detailing the relevant properties of mild steel for its use for the bodywork of cars. Use tables to obtain the relevant data.

Activity

Prepare a report detailing the relevant properties of the polymer ABS for its use for the casing of a telephone.

Problems

1 Determine from tables the following information for materials at about 20°C:

(a) the tensile strength of the carbon steel 1030 in the as rolled condition,
(b) the electrical conductivity on the IACS scale of the unalloyed aluminium 1060 in the annealed state,
(c) the percentage elongation of the brass SCB1,
(d) the yield stress of the manganese steel 120M19 in the quenched and tempered state,
(e) the tensile strength of the stainless steel 302S31 in the soft state,
(f) the density of the plastic ABS,
(g) the plane strain fracture toughness of the plastic polypropylene,
(h) the tensile strength of the elastomer natural rubber,
(i) the thermal expansivity (linear coefficient of expansion) of the plastic high density polythene,
(j) the tensile modulus of the plastic ABS.

2 The rain water guttering used for buildings is required to have a high stiffness per unit weight so that it does not sag under its own weight. Use tables to either obtain the specific modulus of values of the modulus and density and hence compare cast iron, aluminium alloys, and the plastic PVC as possible materials.

3 The panels used for car bodywork need to be in sheet form and stiff. Use tables to obtain modulus of elasticity values and hence compare carbon steel, an aluminium alloy, polypropylene, and a composite formed by polyester with 65% glass fibre cloth.

4 The fan in a vacuum cleaner need to be made of a low density material and a high tensile strength, i.e. a high specific strength. The aluminium alloy LM6 has been suggested because the fan could then be die cast. Use tables to obtain the specific strength of the material.

5 The plastic ABS has been suggested for use as the casing for a radio. The properties required include high stiffness. Determine from tables the modulus of elasticity and compare it with other plastics.

6 The material high tensile brass HTB1 has been suggested as a material for use as a marine propeller. Use tables to obtain values of its tensile strength, 0.1% proof stress and percentage elongation.

7 The alloy steels 150M36 and 530M40 have been suggested as the material for a car axle. Determine from tables the tensile strengths, yield stresses and percentage elongations for both materials, in the quenched and tempered state, so that a comparison can be made.

8 Table 3.2 gives data for cast aluminium alloys when sand cast and in the as manufactured condition. Select, from the list, a material which is likely to be tough.

Table 3.2 *Mechanical properties of cast aluminium alloys*

Material	Tensile strength MPa	% elongation
LM4	140	2
LM5	140	3
LM6	160	5

9 Table 3.3 gives data for polymers. Select, from the list, a material which will be stiff and not too brittle.

Table 3.3 *Mechanical properties of polymers*

Polymer	Tensile strength MPa	Tensile modulus GPa	% elong.
ABS	17–58	1.4–3.1	10–140
Acrylic	50–70	2.7–3.5	5–8
Cellulose acetate	24–65	1.0–2.0	5–55
Cellulose acetate butyrate	18–48	0.5–1.4	40–90
Polyacetal, homopolymer	70	3.6	15–75
Polyamide, Nylon 66	80	2.8–3.3	60–300

10 Table 3.4 gives data for free-cutting steels which have been quenched and tempered to 550-660°C. Select, from the list, the strongest steel.

Table 3.4 *Mechanical properties of free-cutting steels*

Steel	Tensile strength MPa	Yield stress MPa	% elongation
212M36	550–700	340	20
216M36	550–700	380	15
220M44	700–850	450	15

11 Table 3.5 gives data for the electrical resistivities of various coppers and copper alloys at about 20°C. Select from the list the material with the highest conductivity.

Table 3.5 *Electrical properties of copper and copper alloys*

Material	Resistivity 10^6 Ω m
Electrolytic copper, > 99.90% pure	1.71
Oxygen-free copper, > 99.95% pure	1.71
Copper–1% cadmium	2.2
Copper–15% zinc	4.7
Copper–20% zinc	5.4
Copper–2% nickel	5.0
Copper–6% nickel	9.9

12 Table 3.6 gives thermal conductivity values for a number of materials. Which of the materials will feel coldest to touch?

Table 3.6 *Thermal conductivity values*

Material	Thermal conductivity W m^{-1} K^{-1}
Aluminium alloys	120 to 200
Copper alloys	81
Carbon steels	47
Magnesium alloys	80 to 140

4 Materials testing

4.1 Determination of properties

This chapter is a discussion of the standard tests that are used for the determination of the properties of materials. The standard tests used in Britain are those specified by the British Standards Association and European standards organisations. These include:

BSEN 10002	Methods of tensile testing of metals
BS 131	Methods of notched bar tests
BSEN 10045	Charpy test
BS 240	Method for Brinell hardness test
BS 427	Method for Vickers hardness test
BS 891	Method for Rockwell hardness test
BS 1639	Method of bend testing of metals
BS 2782	Methods of testing plastics
BS 4175	Methods for superficial hardness test (Rockwell)
BS 5714	Resistivity measurements for metals

Standards which are specified as BSEN are European standards which have been adopted as British Standards. The above represents just a small selection of the such tests.

4.2 The tensile test

In a tensile test, measurements are made of the force required to extend a standard test piece at a constant rate, the elongation of a specified gauge length of the test piece being measured by some form of extensometer. British and European standards (BSEN 10002 Part 1) state the rate at which the stresses are applied should be between 2 and 10 MPa/s if the tensile modulus is less than 150 GPa and between 6 and 30 MPa/s if the tensile modulus is equal to or greater than 150 GPa. In order to eliminate any variations in tensile test data due to differences in the shapes of test pieces, standard shapes and sizes are adopted.

4.2.1 The test piece

Test pieces are said to be *proportional test pieces* if the relationship between the gauge length L_0 and the cross-sectional area A of the gauge length is:

$$L_0 = k\sqrt{A}$$

British and European standards specify the constant k should have the value 5.65 and the gauge length should be 20 mm or greater. With circular cross-sections of diameter d, $A = \frac{1}{4}\pi d^2$ and thus to a reasonable approximation this value of k gives:

$$L_0 = 5d$$

With circular cross-sectional areas which are too small for this value of k, a higher value may be used, preferably 11.3. When test pieces are proportional test pieces the same test results are given for the same test material when different size test pieces are used.

Figure 4.1 shows the standard size test pieces for round and flat samples of metals with Table 4.1 showing the standard dimensions that can be used. For the tensile test data for the same material to give essentially the same results, regardless of the length of the test piece used, it is vital that the standard dimensions are adhered to. An important feature of the dimensions is the radius given for the shoulders of the test pieces. Very small radii can cause localised stress concentrations which may result in the test piece failing prematurely.

Table 4.1 *Dimensions of standard test pieces*

Flat test pieces

b mm	L_0 mm	L_c mm	L_t mm
20	80	120	140
12.5	50	75	87.5

Round test pieces (proportional)

d mm	A mm^2	L_0 mm	L_c mm
20	314.2	100	110
10	78.5	50	55
5	19.6	25	28

Note: $k = 5.85$.

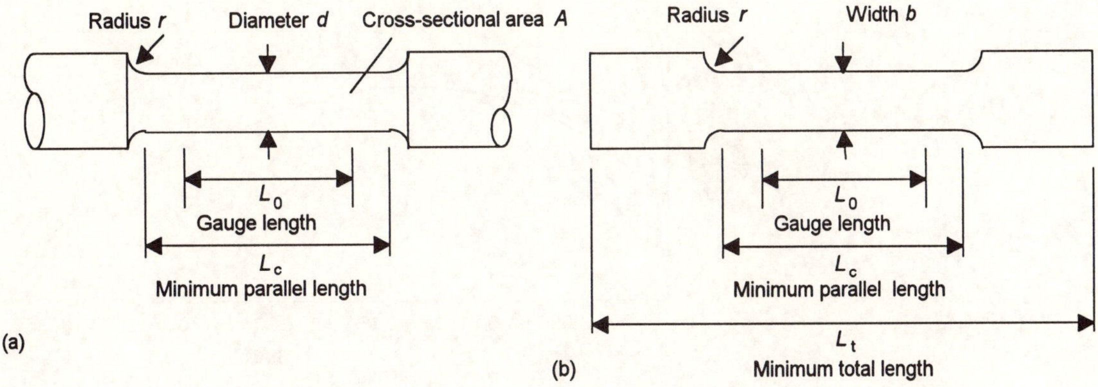

Figure 4.1 *Standard test pieces: (a) round, (b) flat*

4.2.2 Tensile test results

The results of tensile tests are obtained as force-extension data which is generally plotted, either manually or by the machine, as a force-extension graph. Since stress is force/original area and strain is extension/original gauge length then the graph is readily translated into stress-strain. From such graphs the following quantities can be determined (see Chapter 2 for discussion of the terms):

1 The *tensile strength*, this being the stress corresponding to the maximum force.

2 The *yield stress*, this being the stress at which the material begins to yield and show plastic deformation without any increase in load. The term *upper yield stress* is used for the value of the stress when the first decrease in force at the yield is observed and the *lower yield stress* for the lowest value of stress during plastic yielding.

3 The *proof stress*, this being the stress at which the non-proportional extension is equal to a specified percentage of the gauge length. 0.1% and 0.2% are the percentages commonly used.

4 The *tensile modulus*, being the slope of the stress-strain graph over its proportional region.

In addition, measurements of the gauge length before and after breaking enable the *percentage elongation* to be determined, this being permanent elongation of the gauge length expressed as a percentage of the original gauge length.

Example

Figure 4.2 shows a stress-strain graph for a sample of mild steel. Determine (a) the upper yield stress, (b) the lower yield stress and (c) the tensile strength.

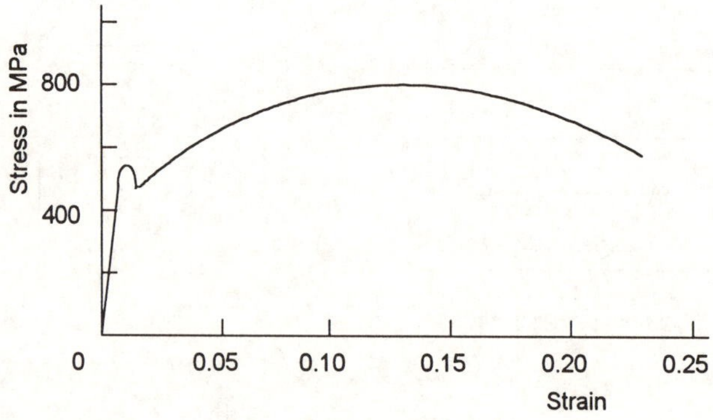

Figure 4.2 *Stress-strain graph for a sample of mild steel*

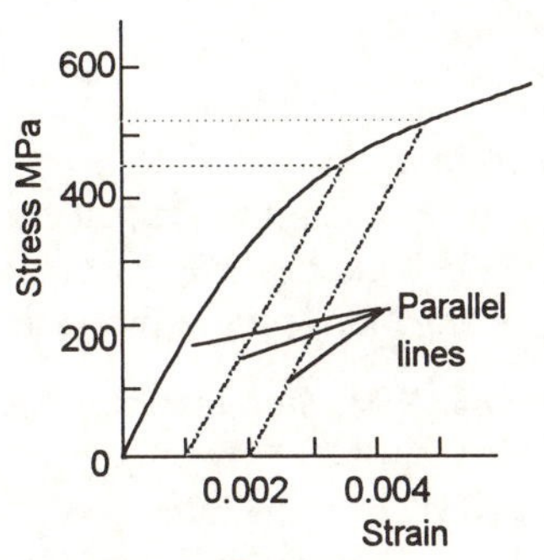

Figure 4.3 *Example*

The upper yield stress is about 550 MPa, the lower yield stress about 500 MPa and the tensile strength about 800 MPa.

Example
Figure 4.3 shows part of the stress-strain graph for a sample of an aluminium alloy. Determine the 0.1% and 0.2% proof stresses.

The 0.1% proof stress is about 450 MPa, the 0.2% about 520 MPa.

4.2.3 Validity of tensile test data

The purpose of taking tensile test pieces and carrying out the tests is to obtain data which enables judgements to be made about the material from which the test piece was cut. The samples of a material have to be taken in such a way that the properties deduced from the tensile test are representative of the material as a whole. There may, however, be problems in assuming this. The following paragraphs outline some of these problems:

1. *The properties of a product may not be the same in all parts of it*
   With a casting there may be different cooling rates in different parts of a casting, e.g. the surface compared with the core, or thin sections compared with thick sections. As a result the internal structure of the material may differ and as a consequence the tensile properties differ. A tensile test piece cut from one part may not thus represent the properties of the entire casting. For the same reason, the properties of a separately cast test piece may not be the same as those of the cast product because the different sizes of the two lead to different cooling rates.

2. *The size of an item affects its properties after heat treatment*
   If the mechanical properties of metals are looked up in tables you will often find that different values of the properties are quoted for different limiting ruling sections. The *limiting ruling section* is the maximum diameter of round bar at the centre of which the specified properties may be obtained. The reason for the difference of mechanical properties of the same material for the different diameter bars is that during the heat treatment different rates of cooling occur at the centres of such bars due to their differences in sizes. Consequently there are differences in microstructure and hence differences in mechanical properties. For example, the steel 070M55 with a limiting ruling section of 19 mm may have tensile strengths of 850 to 1000 MPa, with a limiting ruling section of 63 mm the strengths are 777 to 930 MPa and for a limiting ruling section of 100 mm the strengths are 700 to 830 MPa.

3. *The properties of a product may not be the same in all directions*
   For example, with rolled sheet there is a directionality of properties with the tensile properties in the longitudinal, transverse and through the thickness of the sheet differing. Thus, for example, with

rolled brass strip we might have tensile strengths of 740 MPa in the direction of the rolling and 850 MPa at right angles to it.

4 *The temperature in service of the product may not be the same as that of the test piece when the tensile test data was obtained*
The tensile properties of metals depend on temperature. In general, the tensile modulus and tensile strength both decrease with an increase in temperature, the percentage elongation tends to increase.

5 *The rate of loading of a product may differ from that used with the test piece*
The data obtained from a tensile test is affected by the rate at which the test piece is stretched, so in order to give standardised result the tests are carried out at a constant stress rate, between 2 and 20 MPa/s if the tensile modulus is less than 150 GPa and between 6 and 30 MPa/s if equal to or greater than 150 GPa.

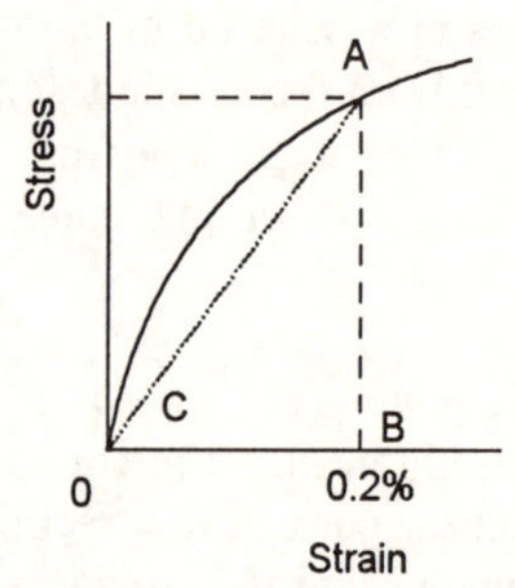

Figure 4.4 *The secant modulus is AB/BC*

4.2.4 Interpreting tensile test data

The results from tensile tests can be used to determine the safe stresses to which a material can be subject. Thus, the higher the yield stress of a metal the higher the stresses that it can be exposed to in service without yielding. Another important deduction that can be made is whether the material is brittle or ductile. A brittle material will show little plastic behaviour and have a low percentage elongation. A ductile material will shows considerable plastic behaviour and have a high percentage elongation.

A consequence of the heat treatment and working of a material that occur during the fabrication of products is a change in mechanical properties. Thus tensile test data enable the effectiveness of heat treatments and the effects of working to be monitored.

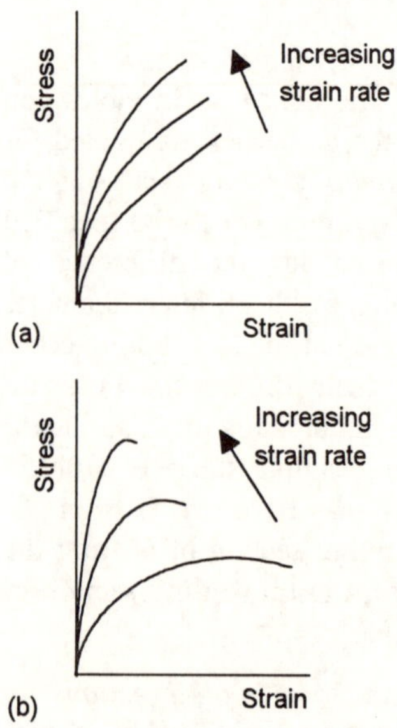

Figure 4.5 *(a) Brittle, (b) ductile plastic*

4.2.5 Tensile tests for plastics

Tensile tests can be used with plastic test pieces to obtain stress-strain data. The term tensile strength has the same meaning as with metals. However, the tensile modulus, i.e. the slope of the stress-strain graph over the proportional region, cannot always be easily obtained. For many plastics there is no really straight-line part of the stress-strain graph. Thus, as a measure of the stiffness of the material, a modulus is defined in a different way. The *secant modulus* is obtained by dividing the stress at a strain of 0.2% by that strain, as illustrated in Figure 4.4.

The stress-strain properties of plastics are much more dependent than metals on the rate at which the strain is applied. Thus, for example, the tensile test may indicate a yield stress of 62 MPa when the rate of elongation is 12.5 mm/min but 74 MPa when it is 50 mm/min. Also the form of the stress-strain graph may change with a ductile material at low strain rates becoming a brittle one at high strain rates. Figure 4.5 shows the general forms of stress-strain graphs for plastics at different strain rates. Another factor that is more marked than with metals is the effect of temperature on the properties of plastics.

Example

Figure 4.6 is the stress-strain graph for a sample of ABS Novodur grade PK (courtesy of Bayer (UK) Ltd.). Estimate (a) the tensile modulus and (b) the tensile strength.

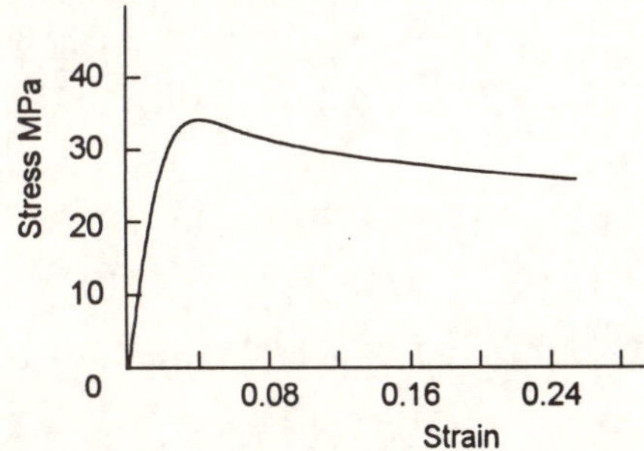

Figure 4.6 *Stress–strain graph for Novodur PK (adapted from Bayer (UK) Ltd.)*

(a) The tensile modulus is the slope of the proportional part of the stress-strain graph and is thus about 28/0.02 = 1400 MPa = 1.4 GPa.
(b) The tensile strength is the maximum stress. This is about 34 MPa.

4.2.6 Verification of tensile test equipment

The British and European Standard BSEN 10002 Part 2 describes how the force readings given by such a machine can be verified. A given force indicated by the machine is compared with the true force indicated by a force proving instrument or exerted by masses. Three series of measurements should be taken with increasing force, each series having at least five steps at regular intervals from 20% of the maximum range of the scale.

Activity

Carry out a tensile test of a sample test piece and present your results.

Activity

Carry out the following simple experiments to obtain information about the tensile properties of materials when commercially made tensile testing equipment is not available. *Safety note*: when doing experiments involving the stretching of wires, filaments, glass fibres or other materials, the specimen may fly up into your face when it breaks. When a taut wire snaps, a lot of stored elastic energy is suddenly released. *Safety spectacles should be worn.*

Obtain a force–extension graph for rubber by hanging a rubber band (e.g. 74 mm by 3 mm by 1 mm band) over a clamp or other fixture, adding masses to a hanger suspended from it and measuring the extension with a ruler (Figure 4.7).

Obtain a force–extension graph for a nylon fishing line in a similar way, the fishing line being tied to form a loop (e.g. about 75 cm long).

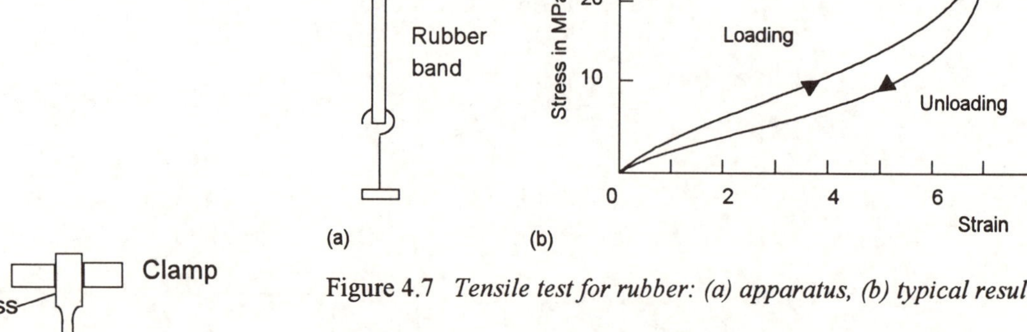

Figure 4.7 *Tensile test for rubber: (a) apparatus, (b) typical result*

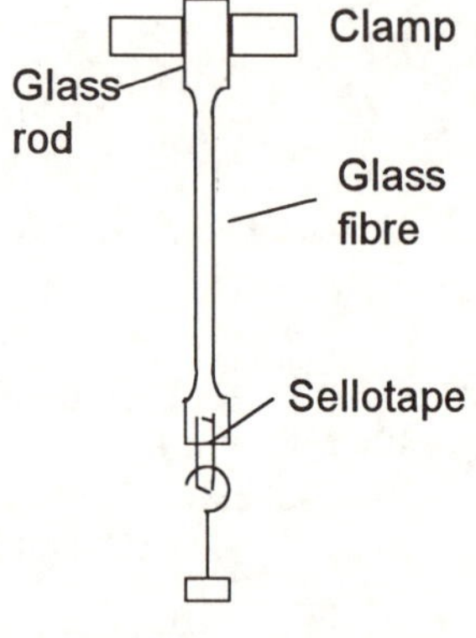

Figure 4.8 *Tensile test for glass*

Carry out a tensile testing of glass. A glass fibre for tensile testing can be obtained if the middle of a glass rod (e.g. about 3 mm diameter and 20 cm long) is softened in a Bunsen flame and drawn out into a fibre. The top of the glass rod can be clamped and a hanger attached to the lower end by means of Sellotape (Figure 4.8) or the lower end of the glass rod being formed into hook.

Carry out a tensile testing of a metal wire. Figure 4.9 shows how the force–extension graph, and hence the tensile modulus of elasticity, can be determined for a metal wire (e.g. iron wire with a diameter of about 0.2 mm, copper wire about 0.3 mm diameter, steel wire about 0.08 mm diameter, all having lengths of about 2.0 m). The initial diameter d of the wire is measured using a micrometer screw gauge. The length L of the wire from the clamped end to the marker (a strip of paper attached by Sellotape) is measured by using a rule, a small load being used to give a taut wire. Masses are then added to the hanger and the change in length e from the initial position recorded. Hence data can be obtained to plot a graph of force (F) against extension (e).

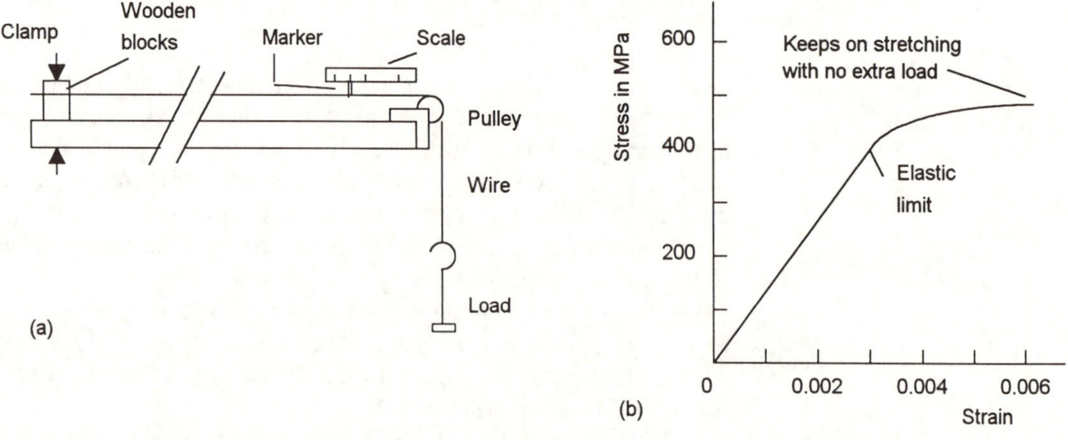

Figure 4.9 *Tensile test for wire: (a) apparatus, (b) example of result with copper wire*

Determine the tensile modulus of elasticity E, i.e. the gradient F/e of the initial straight-line part of the graph.

$$E = \frac{\text{stress}}{\text{strain}} = \frac{\left(F/\frac{1}{4}\pi d^2\right)}{e/L} = \frac{F}{e} \times \frac{4L}{\pi d^2}$$

Example

A length of wire was subject to a tensile test using the apparatus shown in Figure 4.10. The length of wire from the clamp to the marker was 2.0 m and the diameter of the wire 0.08 mm. Measurements were made of the movements of the marker as loads were added to the wire. A graph of load plotted against marker movement gave a straight-line graph through the origin with a slope of 0.051 kg/mm. What is the tensile modulus?

Using the equation derived above,

$$E = 0.051 \times 1000 \times 9.8 \times \frac{4 \times 2}{\pi \times (0.08 \times 10^{-3})^2}$$

$$= 199 \times 10^9 \text{ Pa} = 199 \text{ GPa}$$

4.3 Bend tests

A simple test that is often quoted by suppliers of materials as a measure of ductility is the *bend test*. The test involves bending a sample of the material through some angle and determining whether the material is unbroken and free from cracks after the bending. There are a number of ways that can be used to carry out such a test, BS 1639 listing the British Standards, Figure 4.10(a), (b) and (c) giving commonly used methods.

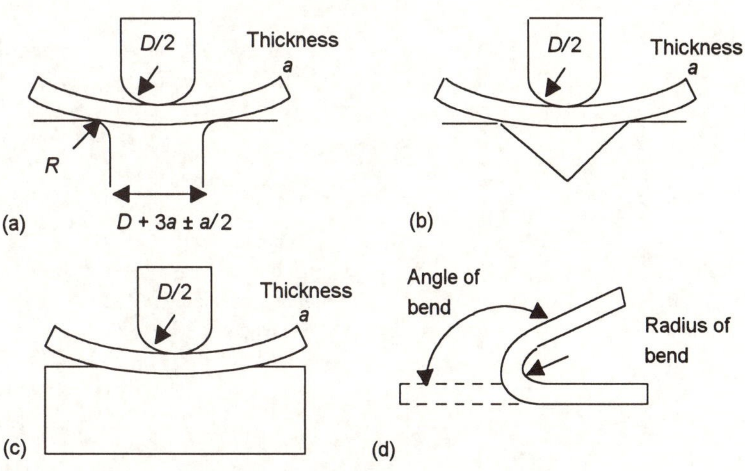

Figure 4.10 *Bend test: (a) mandrel test, (b) bending on a vee block, (c) bending on a block of soft material, (d) the angle of bend*

The simplest method is the mandrel form of test shown in Figure 4.10(a), this being suitable for medium and thin thickness sheet for angles of bend up to 120°. Figure 4.10(b) shows how the test can be conducted on a vee block, this being suitable for medium thickness sheet with bend angles up to 90°. Figure 4.10(c) shows the form of test possible for thin sheet with bend angles up to 90°, the material being bent on a block of soft material. Other methods can also be used, e.g. bending round a mandrel, free bending and pressure bending (see the British Standard for more details). The results of a bend test are quoted in terms of the angle of bend that can be withstood without breaking or cracking, as illustrated in Figure 4.10(d).

Activity
Carry out a bend test on a strip of aluminium alloy. Make up the apparatus for a mandrel test.

4.4 Impact tests

Impact tests are designed to simulate the response of a material to a high rate of loading and involve a test piece being struck a sudden blow. There are two main forms of test, the *Izod* and *Charpy* tests. Both tests involve the same type of measurement but differ in the form of the test pieces. Both involve a pendulum (Figure 4.11) swinging down from a specified height h_0 to hit the test piece and fracture it. The height h to which the pendulum rises after striking and breaking the test piece is a measure of the energy used in the breaking. If no energy were used the pendulum would swing up to the same height h_0 it started from, i.e. the potential energy mgh_0 at the top of the pendulum swing before and after the collision would be the same. The greater the energy used in the breaking, the greater the 'loss' of energy and so the lower the height to which the pendulum rises. If the pendulum swings up to a height h after breaking the test piece then the energy used to break it is $mgh_0 - mgh$.

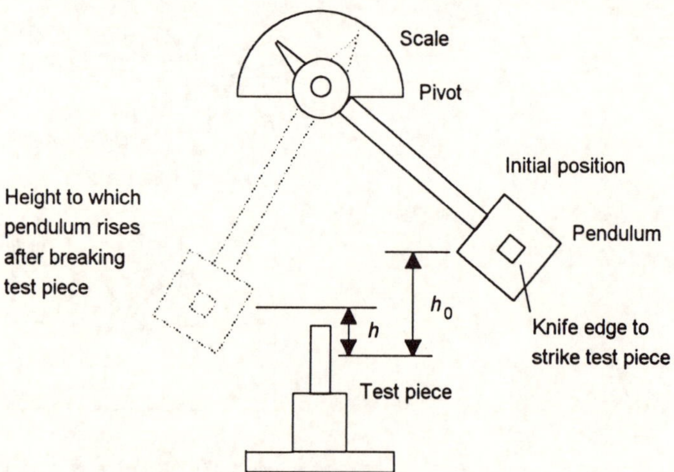

Figure 4.11 *Impact testing*

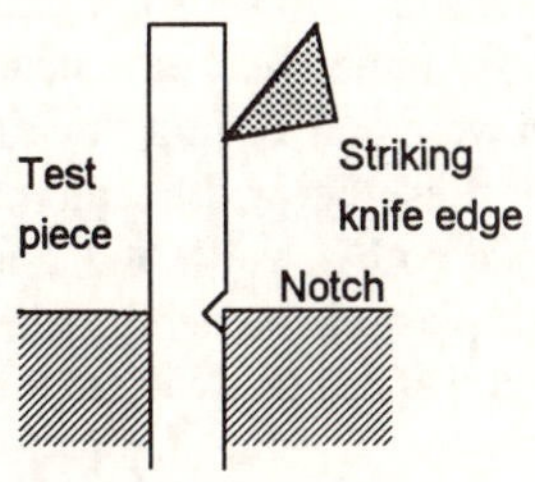

Figure 4.12 *Izod form of test*

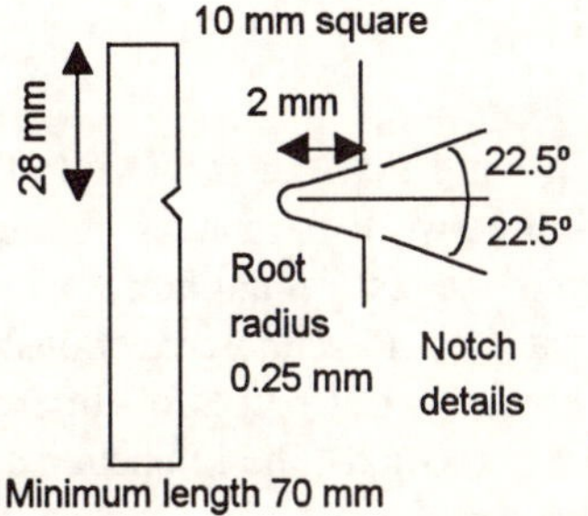

Figure 4.13 *Izod metal test piece*

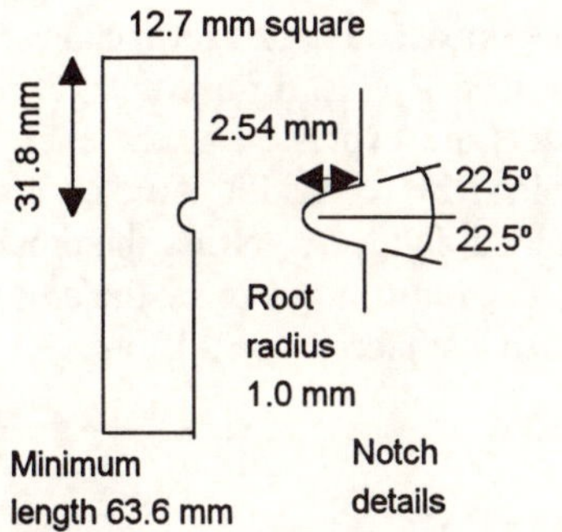

Figure 4.14 *Izod plastic test piece*

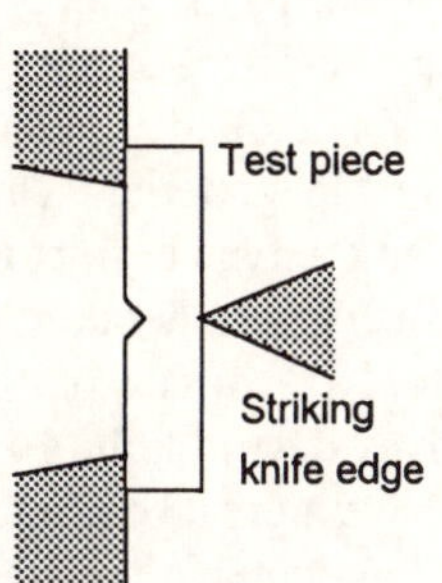

Figure 4.15 *Charpy test*

4.4.1 Izod test pieces

With the Izod test the energy absorbed in breaking a cantilevered test piece is measured, as illustrated by Figure 4.12. The test piece has a notch and the blow is struck on the same face as the notch and at a fixed height above it. In the case of metals, the test pieces used are generally either 10 mm square or 11.4 mm in diameter if round. Figure 4.13 shows details of one form of the square test piece. With the 70 mm length the notch is 28 mm from the top of the piece. If a longer length is used then more than one notch is used. With a length of 96 mm there are two notches on opposite faces, one 28 mm from the top and the other twice that distance from the top. With a longer length test piece of 126 mm there are three notches, on three of the faces. The first notch is 28 mm from the top, the second twice that distance and third three times that distance from the top.

In the case of plastics, the test pieces are 12.7 mm square or 12.7 mm by 6.4 to 12.7 mm depending on the thickness of the material concerned. Figure 4.14 shows details of such a test piece. With metals the pendulum strikes the test piece with a speed of between 3 and 4 m/s, with plastics a lower speed of 2.44 m/s is used.

4.4.2 Charpy test pieces

With the Charpy test, the energy absorbed in breaking a test piece in the form of a beam is measured (Figure 4.15). The standard machine has the pendulum hitting the test piece with an energy of 300 ± 10 J. The test piece is supported at each end and notched at the midpoint between the two supports. The notch is on the face directly opposite to where the pendulum strikes the test piece. The British and European Standard is BSEN 10045.

For metals, the test piece generally has a square cross-section of side 10 mm and length 55 mm with there being 40 mm between the supports. Figure 4.16 shows details of such a test piece and the forms of notch commonly used.

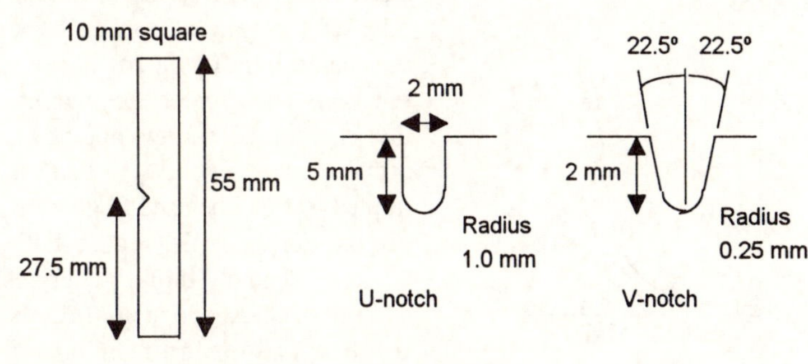

Figure 4.16 *Charpy metal test pieces*

With the V-notch, reduced width specimens of 7.5 mm and 5 mm can be used. For plastics, the test pieces may be unnotched or notched. A standard test piece is 120 mm long, 15 mm wide and 10 mm thick in the case of moulded plastics. With sheet plastics the width can be the thickness of the sheet. The notch is U-shaped with a width of 2 mm and a radius of 0.2 mm at its base. For moulded plastics the depth below the notch is 6.7 mm, for the sheet plastics either 10 mm or two-thirds of the sheet thickness.

Activity
Carry out an impact test using commercial apparatus.

4.4.3 Impact test results

In stating the results of impact tests it is vital that the form of test is specified. There is no reliable relationship between the values obtained by the two forms of test and so values from one test cannot be compared with those from the other. In addition there is no reliable relationship between the impact energies given for breaking test pieces of different sizes or different notches with the same test method. The impact energy value obtained for a material is influenced by such factors as the temperature, the speed of impact, any degree of directionality occuring in the properties of the material from which the test piece was cut, and the thickness of the test piece.

For both the Izod and Charpy tests, the impact strengths for metals are expressed in the form of the energy absorbed, i.e. as, for example, 30 J. For plastics, with the Izod test the results are expressed as the energy absorbed in breaking the test piece divided by the width of notch and with the Charpy test as the energy absorbed divided by either the cross-sectional area of the specimen for unnotched test pieces or by the cross-sectional area behind the notch for notched test pieces, e.g. 2 kJ/m^2.

4.4.4 Interpreting impact test results

When a material is stretched energy is stored in the material. Think of stretching a spring or a rubber band. When the stretching force is released the material springs back and the energy is released. However, if the material suffers a permanent deformation then all the energy is not released. The greater the amount of such plastic deformation the greater the amount of energy not released. Thus when a ductile material is broken, more energy is 'lost' than with a brittle material. Thus the impact test can be used to give information about the type of fracture that occurs. For example, Figure 4.17 shows the effect of temperature on the Charpy V-notch impact energies obtained for test pieces of a 0.2% carbon steel. Above about 0°C the material gives ductile failures, below that temperature, brittle failures. Such graphs have a great bearing on the use that can be made of the material, since at low temperatures the steel can be easily shattered by impact. Table 4.2 shows some typical impact strengths for metals.

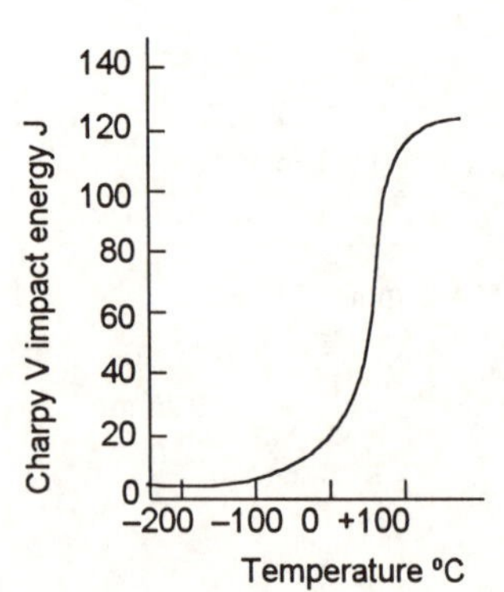

Figure 4.17 *Effect of temperature for a 0.2% carbon steel*

Table 4.2 *Impact strengths at room temperature for metals*

Materials	Charpy V impact strength J
Aluminium, commercially pure, annealed	30
Aluminium–1.5% Mn alloy, annealed	80
hard	34
Copper, oxygen-free HC, annealed	70
Cartridge brass (70% Cu, 30% Zn), annealed	88
¾ hard	21
Cupronickel (70% Cu, 30% Ni), annealed	157
Magnesium–3% Al, 1% Zn alloy, annealed	8
Nickel alloy, Monel, annealed	290
Titanium–5% Al, 2.5% Sn, annealed	24
Grey cast iron	3
Malleable cast iron, Blackheart, annealed	15
Austenitic stainless steel, annealed	217
Carbon steel, 0.2% carbon, as rolled	50

The appearance of the fractured surfaces after an impact test also gives information about the type of fracture that has occurred. With a brittle fracture of metals, the surfaces are crystalline in appearance. With a ductile fracture, the surfaces are rough and fibrous in appearance. Also with ductile failure there is a significant reduction in the cross-sectional area of the test piece, but with brittle fracture there is virtually no such change.

Table 4.3 gives some typical values of impact strengths for plastics at 20°C for the Izod test with a notch tip having a radius of 0.25 mm and a depth of 2.75 mm. The impact properties of plastics vary quite significantly with temperature, changing in many cases from brittle to tough at some particular transition temperature.

Table 4.3 *Impact strengths for plastics*

Material	Impact strength kJ/m^2
Polythene, high density	30
ABS	25
Nylon 6.6, dry	5
Polyvinyl chloride, unplasticised	3
Polystyrene	2

With plastics, a brittle failure gives fracture surfaces which are smooth and glassy or somewhat splintered, with a ductile failure the surfaces often have a whitened appearance. Also, the change in cross-sectional area can be considerable with a ductile failure but negligible with brittle failure.

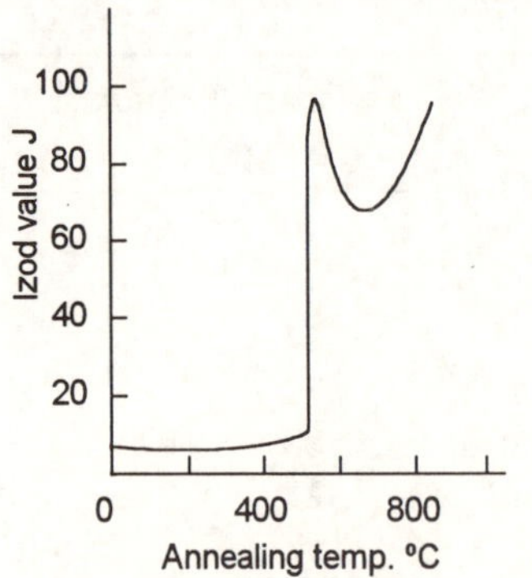

Figure 4.18 *Effect of annealing temperature on Izod values*

One use of impact tests is to determine whether heat treatment has been successfully carried out. A comparatively small change in heat treatment can lead to quite noticeable changes in impact test results. The changes can be more pronounced than changes in other mechanical properties such as percentage elongation or tensile strength. Figure 4.18 shows the effect on the Izod impact test results for cold-worked mild steel of annealing to different temperatures. The impact test can thus be used to indicate whether annealing has been carried out to the required temperature.

Example

A sample of unplasticised PVC has an impact strength of 3 kJ/m^2 at 20°C and 10 kJ/m^2 at 40°C. Is the material becoming more or less brittle as the temperature is increased?

Because there is an increase in the impact energy the material is becoming more ductile.

4.5 Fracture toughness tests

Fracture toughness testing involves test pieces with sharp notches being strained until the crack propagates and the test piece fails. The problem in obtaining test pieces is producing the sharp notches. This is done by taking a test piece with a machined notch and then using a standardised pre-cracking procedure involving loading with an alternating stress (fatigue loading) in order to obtain a sharp crack, length a, at the base of the machined notch. Figure 4.19 shows forms of test piece, as specified by BS 7448. For the test, a steadily increasing force is applied to a test piece, the notch opening as a result, and the maximum value of the force recorded before breaking occurs as a result of the crack propagating.

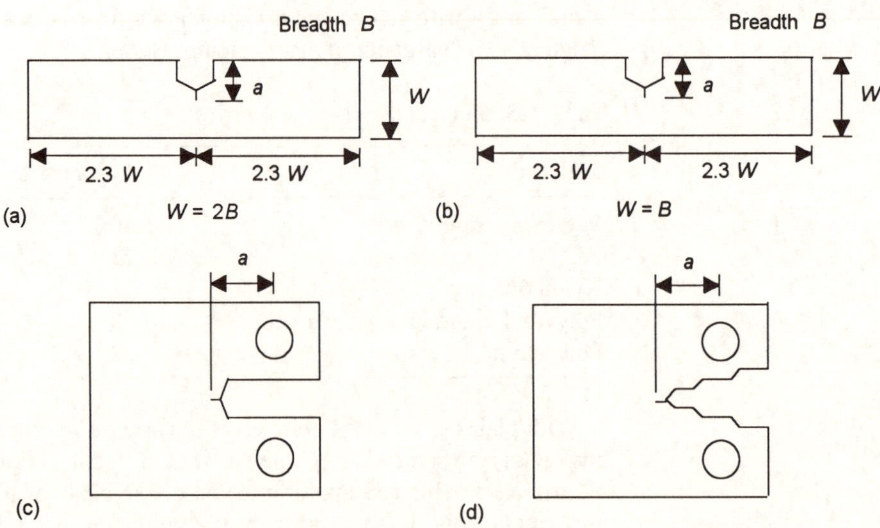

Figure 4.19 *Fracture toughness test pieces: (a) and (b) for three-point bending, (c) and (d) for tensile loading*

4.6 Hardness tests

The hardness of a material may be specified in terms of some standard test involving indenting or scratching of the surface of the material, the harder a material the more difficult it is to make an indentation or scratch. There is no absolute scale for hardness, each hardness form of test having its own scale. Though some relationships exist between results on one scale and those on another, care has to be taken in making comparisons because the different types of test are measuring different things.

The most common form of hardness tests for metals involves standard indentors being pressed into the surface of the material concerned. Measurements associated with the indentation are then taken as a measure of the hardness of the surface. The Brinell test, the Vickers test and the Rockwell test are the main forms of such tests.

4.6.1 The Brinell test

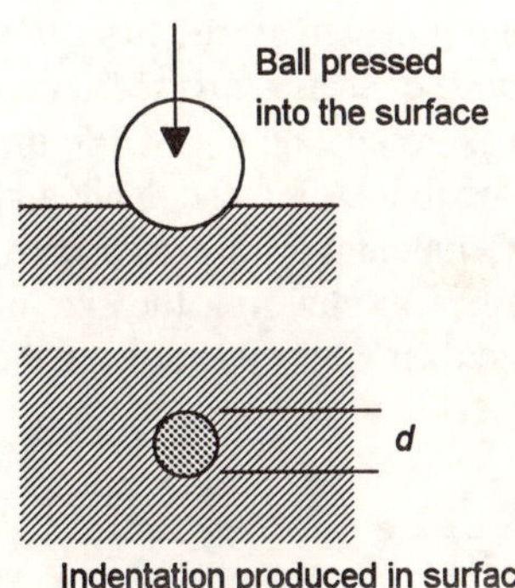

Figure 4.20 *Brinell test*

With the Brinell test, a hardened steel ball is pressed for a time of 10 to 15 s into the surface of the material by a standard force (Figure 4.20). After the load and ball have been removed, the diameter of the indentation is measured. The Brinell hardness number, signified by HB, is obtained by dividing the size of the force applied by the surface area of the spherical indentation:

$$\text{Brinell hardness number} = \frac{\text{applied force}}{\text{surface area of indentation}}$$

The units used for the area are mm^2 and for the force kgf (1 kgf = 9.8 N and is the gravitational force exerted by 1 kg). The area can be obtained, from the measured diameter of the indentation and ball diameter, either by calculation or the use of tables:

$$\text{area} = \tfrac{1}{2}\pi D\left[D - \sqrt{D^2 - d^2}\,\right]$$

where D is the diameter of the ball and d that of the indentation.

The diameter D (mm) of the ball used and the size of the applied force F (kgf units) are chosen, for the British Standard, to give F/D^2 values of 1, 5, 10 or 30 with the diameters of the balls being 1, 2, 5 or 10 mm. In principle, the same value of F/D^2 should give the same hardness value, regardless of the diameter of the ball used. It is necessary for the impression to have a diameter of between $0.25D$ and $0.50D$ if accurate values of the hardness are to be obtained. The F/D^2 value is thus chosen to fit the materials concerned, the harder the material the higher the value used. For steels and cast iron the value used for F/D^2 is 30, for copper alloys and aluminium alloys 10, for pure copper and aluminium 5 and for lead, tin and tin alloys 1.

The Brinell test cannot be used with very soft or very hard materials. In the one case the indentation becomes equal to the diameter of the ball and in the other there is either no or little indentation on which measurements can be based. Also, with very hard materials the material deforms the indenter. The Brinell test is thus limited to materials with hardnesses up to about 450 HB with a steel ball and 600 HB with a tungsten carbide ball.

The thickness of the material being tested should be at least ten times the depth of the indentation if the results are not to be affected by the thickness of the material. Also, because of the large depth of the indentation, it cannot be used on plated or surface hardened materials since the result will be affected by the underlying material.

Example

For a Brinell test with a steel, what load should be used with a 10 mm diameter ball?

For steels F/D^2 is taken as 30 and thus $F = 30 \times 10^2 = 3000$ kg.

4.6.2 The Vickers test

The Vickers hardness test involves a diamond indenter, in the form of a square-based pyramid with an apex angle of 136°, being pressed under load for 10 to 15 s into the surface of the material under test (Figure 4.21). The result is a square-shaped impression. After the load and indenter are removed the diagonals d of the indentation are measured. The Vickers hardness number (HV) is obtained by dividing the size of the force F, in units of kgf, applied by the surface area, in mm², of the indentation:

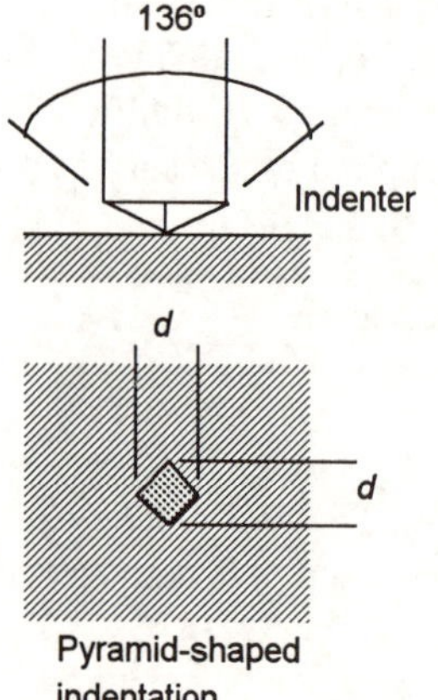

Figure 4.21 *Vickers hardness test*

$$\text{Vickers hardness} = \frac{\text{applied force}}{\text{surface area of indentation}}$$

The surface area can be calculated from the mean diagonal value, the indentation being assumed to be a right pyramid with a square base and an apex angle θ of 136°, or obtained by using tables:

$$\text{Area} = \frac{d^2}{2 \sin \theta/2} = \frac{d^2}{1.854}$$

Thus the Vickers hardness HV is given by:

$$\text{HV} = \frac{1.854F}{d^2}$$

The Vickers test has the advantage over the Brinell test of the increased accuracy that is possible in determining the diagonals of a square as opposed to the diameter of a pyramid. The square indentation produced for a particular material depends on the force used, but because the indentation is always the same square shape regardless of how big the force and the surface area is proportional to the force, the hardness value obtained is independent of the size of the force used. Typically a load of 30 kg is used for steels and cast irons, 10 kg for copper alloys, 5 kg for pure copper and aluminium alloys, 2.5 kg for pure aluminium and 1 kg for lead, tin and tin alloys. Up to a hardness value of about 300, the hardness value number given by the Vickers test is the same as that given by the Brinell test.

4.6.3 The Rockwell test

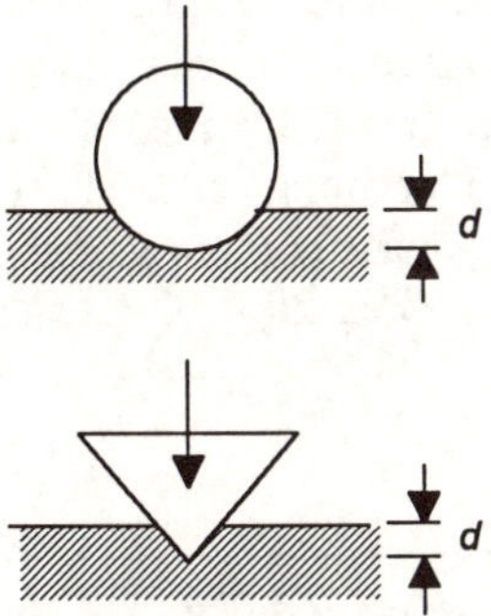

Figure 4.22 *Rockwell test*

The Rockwell hardness test differs from the Brinell and Vickers hardness tests in not obtaining a value for the hardness in terms of the area of an indentation but using the depth of indentation, this depth being directly indicated by a pointer on a calibrated scale. The test uses either a diamond cone or a hardened steel ball as the indenter (Figure 4.22). A preliminary force is applied to press the indenter into contact with the surface. A further force is then applied and causes the indenter to penetrate into the material. The additional force is then removed and there is some reduction in the depth of the indenter due to the deformation of the material not being entirely plastic. The difference in the final depth of the indenter and the initial depth, before the additional force was applied, is determined. This is the permanent increase in penetration e due to the additional force. The Rockwell hardness number is then given by:

$$\text{Rockwell hardness number (HR)} = E - e$$

where E is a constant determined by the form of the indenter. For the diamond cone indenter E is 100, for the steel ball 130.

There are a number of Rockwell scales (Table 4.4), the scale being determined by the indenter and the additional force used. In any reference to the results of a Rockwell test the scale letter must be quoted. For metals the B and C scales are probably the most commonly used ones.

Table 4.4 *Rockwell hardness scales*

Scale	Indenter	Additional load kg	Typical applications
A	Diamond	60	Extremely hard materials, e.g. tool steels
B	Ball 1.588 mm dia.	100	Softer materials, e.g. Cu alloys, Al alloys, mild steel
C	Diamond	150	Hard materials, e.g. steels, hard cast irons, alloy steels
D	Diamond	100	Medium case hardened materials
E	Ball 3.175 mm dia.	100	Soft materials, e.g. Al alloys, Mg alloys, bearing metals
F	Ball 1.588 mm dia.	60	As E, the smaller ball being more appropriate where inhomogeneities exist
G	Ball 1.588 mm dia.	150	Malleable irons, gun metals, bronzes
H	Ball 3.175 mm dia.	60	Soft aluminium, lead, zinc, thermoplastics
K	Ball 3.175 mm dia.	150	Aluminium and magnesium alloys
L	Ball 6.350 mm dia.	60	Soft thermoplastics
M	Ball 6.350 mm dia.	100	Thermoplastics
R	Ball 12.70 mm dia.	60	Very soft thermoplastics

Note: the diameter of the balls arise from standard sizes in inches, 1.588 mm being 1/16 in, 3.175 mm being 1/8 in, 6.350 mm being 1/4 in, and 12.70 mm being 1/2 in.

For the most commonly used indenters with the Rockwell test the size of the indentation is rather small. Localised variations of structure, composition and roughness can thus affect the results. The Rockwell test is, however, more suitable for workshop or 'on-site' use as it less affected by surface conditions than the Brinell or Vickers tests which require flat and polished surfaces to permit accurate measurements. A variation of the Rockwell test has to be used for thin sheet, this test being referred to as the *Rockwell superficial hardness test*. Smaller forces are used and the depth of indentation which is correspondingly smaller is measured with a more sensitive device. The initial force used is 29.4 N. Table 4.5 indicates the scales given by this test.

Table 4.5 *Rockwell superficial hardness scales*

Scale	*Indenter*	*Additional load kg*
15-N	Diamond	15
30-N	Diamond	30
45-N	Diamond	45
15-T	Ball 1.588 mm dia.	15
30-T	Ball 1.588 mm dia.	30
45-T	Ball 1.588 mm dia.	45

Note: the N scales are used for materials that if thick enough would have been tested on the C scale, the T scales for those on the B scale.

Activity
Carry out a hardness measurement using commercial apparatus.

Activity
In the absence of commercial apparatus, devise a simple method of comparing hardnesses of different materials by dropping a punch vertically onto surfaces.

4.6.4 Comparison of the different hardness scales

The Brinell and Vickers tests both involve measurements of the surface area of indentations, the forms of the indenters used being different. The Rockwell test involves measurements of the depth of penetration of indenters. Thus the various tests are concerned with different measurements as an indication of hardness. Consequently the values given by the different methods differ for the same material. There are no simple theoretical relationships between the various hardness scales, though some simple approximate, experimentally derived, relationships have been obtained. Different relationships, however, hold for different metals. The relationships are often presented in the form of tables. Table 4.6 shows part of a table for steels. Up to a hardness value of 300 the Vickers and Brinell values are almost identical.

There is an approximate relationship between hardness values and tensile strengths:

Table 4.6 *Comparison of hardness scales for steels*

Brinell value	Vickers value	Rockwell B	Rockwell C
112	114	66	
121	121	70	
131	137	74	
140	148	78	
153	162	82	
166	175	86	4
174	182	88	7
183	192	90	9
192	202	92	12
202	213	94	14
210	222	96	17
228	240	98	20
248	248	102	24
262	263	103	26
285	287	105	30
302	305	107	32
321	327	108	34
341	350	109	36
370	392		40
390	412		42
410	435		44
431	459		46
452	485		48
475	510		50
500	545		52

$$\text{tensile strength} = k \times \text{hardness}$$

where k is a constant for a particular material. Thus for annealed steels the tensile strength in MPa is about 3.54 times the Brinell hardness value, and for quenched and tempered steels 3.24 times the Brinell hardness value. For brass the factor is about 5.6 and for aluminium alloys about 4.2.

Example

An aluminium alloy has a hardness of 45 HB when annealed and 100 HB when solution treated and precipitation hardened. Estimate the tensile strengths of the alloys in these two conditions if a factor of 4.2 is assumed.

Using a factor of 4.2, then the tensile strength in the annealed condition is $4.2 \times 45 = 189$ MPa. For the heat-treated condition it is $4.2 \times 100 = 420$ MPa. The measured values were 180 MPa and 430 MPa.

4.6.5 Hardness measurements with plastics

The Brinell, Vickers and Rockwell tests can be used with plastics. The Rockwell test with its measurement of penetration depth, rather than surface area, is more widely used. Scale R is a commonly used scale.

Another test that is used with plastics involves an indenter, a ball of diameter 2.38 mm, being pressed against the plastic by an initial force of 0.294 N for 5 s and then an additional force of 5.25 N being applied for 30 s. The difference between the two penetration depths is measured and expressed as a *softness number*. This is just the depth expressed in units of 0.01 mm. Thus a difference in penetration of 0.05 mm is a softness number of 5. The test is carried out at a temperature of 23 ± 1°C.

Another form of test that is used is the *Shore durometer*. This is a hand-held device which involves a rounded indenter being pressed into the surface of the material under the action of a spring or weight, a pointer then registering the hardness value on a scale. A number of scales are used, ranging from Shore A for the very soft to Shore D for the very hard.

4.6.6 The Moh scale of hardness

A completely different form of hardness test, called the Moh scale, is based on assessing the resistance of a material to being scratched. Ten styli with points made of different materials are used. The styli materials are arranged in a scale so that each one will scratch the one preceding it in the scale but not the one that follows it. The scale and materials are:

1 Talc
2 Gypsum
3 Calcspar
4 Fluorspar
5 Apatite
6 Felspar
7 Quartz
8 Topas
9 Corundum
10 Diamond

Thus, for example, felspar will scratch apatite but not quartz. Diamond will scratch all the materials while talc will scratch none of them. In this test the various styli are used until the lowest number stylus is found that will just scratch it. The hardness number is then one less since it is the number of the stylus that just fails to scratch the material. For example, glass can just be scratched by felspar but not by apatite. The glass thus has a hardness number of 5.

4.7 Electrical measurements

Determination of the electrical resistivity or conductivity of a material requires a measurement of the resistance of a strip or block of the material. The British Standard for resistivity measurements with metals

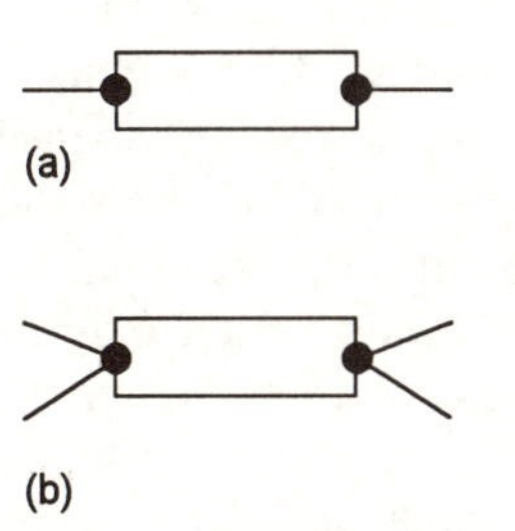

Figure 4.23 *(a) Two-terminal, (b) four terminal*

is BS 5714. In the case of metals the resistivity is very low and so the resistance to be measured can be low. For example, the resistance of a 1 m length of copper wire with a diameter of 1 mm is about 0.03 Ω at 20°C. Such a resistance is not easy to measure, since the means by which it is connected to the measurement system can have resistances of the same order of size or even larger. A smaller gauge wire of 0.1 mm gives a resistance of about 2.1 Ω and is easier to measure. For routine measurements with resistances greater than 1 Ω the test piece can be what is termed a *two-terminal device*, i.e. there is just a single terminal at each end of the test piece for connections (Figure 4.23(a)). For resistances less than 1 Ω the test piece should be a *four-terminal device*, i.e. there are two terminals at each end (Figure 4.23(b)). This means, since measurements require more than one connection to each end of a resistor, that the circuit connections to each end give less ambiguity as to between which points measurements are being made. For routine resistance measurements the method used should be capable of an accuracy of at least ±0.30%. The method used for such resistance measurements is likely to be a resistance bridge, with possibly a Kelvin double bridge for small resistances.

In addition to measuring the resistance, the length and cross-sectional area of the test piece is required. The area can be obtained by direct measurement; however, an alternative method which is often used is to weigh the test piece and calculate the area from a knowledge of the density and length, the area being mass/(density × length).

Since resistivity changes with temperature, it is important that the temperature t at which a measurement is made is noted. The following equation can then be used to correct the result to the reference temperature t_0 at which the result is required.

$$\rho_{t_0} = \frac{\rho_t}{1 + (\alpha + \gamma)(t - t_0)}$$

where α is the temperature coefficient of resistance at the reference temperature and γ the coefficient of linear expansion.

For materials such as plastics or ceramics, the problem is that they have very high resistivities. This can present the problem that the surface layers, perhaps as a result of the absorption of moisture, might have a significantly lower resistivity than the bulk of the material and so the value indicated by the measurement is not that of the bulk material. Polymers also present the problem that when a voltage is applied across a sample that the current through the material slowly decreases with time. Thus resistivity measurements need to have a time quoted with them, e.g. the value one minute after the application of a voltage.

Activity

Use the *ammeter–voltmeter method* for the measurement of resistance of a length of wire and hence obtain its resistivity. Connect an ammeter in series with the length of wire and a voltmeter in parallel with it (Figure 4.24). Take readings of the current and voltage for different input voltages to the circuit and plot a graph of the voltage across the

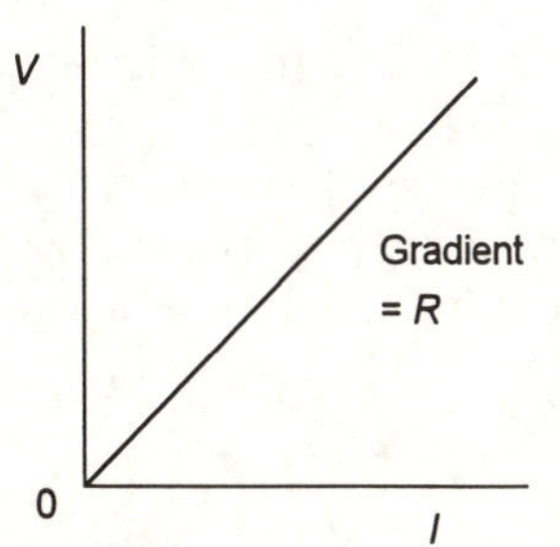

Figure 4.24 *Ammeter–voltmeter method*

resistance wire against the current through it. The gradient of the graph is the resistance R of the length of wire. To obtain the resistivity, the length of the wire and its diameter have to be measured. The diameter should be measured at a number of points and an average value obtained.

The accuracy of the ammeter–voltmeter method of resistance measurement depends on the accuracy of the meters used and the resistance of the voltmeter used. The ammeter measures not only the current through the resistance wire but also the current through the voltmeter. The higher the resistance of the voltmeter relative to that of the resistance being measured, the smaller the fraction of the current that passes through the voltmeter and so the closer the reading of the ammeter is to the current through the resistance. For the measurement of resistances of a few ohms with a voltmeter with resistance of thousands of ohms, the effect on the accuracy is likely to be considerably smaller than the limitations imposed by the accuracy with which the ammeter and voltmeter can be read. With ammeters and voltmeters having pointers moving across scales, the accuracy with which readings can be made is likely to be of the order of a few per cent. For example, with a voltmeter with a scale of 0 to 1 V and scale divisions of 0.1 V we might consider that we can estimate the position of the pointer setting to within ±0.2 of a scale division. This is an accuracy of ±0.02 V. In a full-scale reading of 1 V this would represent an accuracy of ±2%.

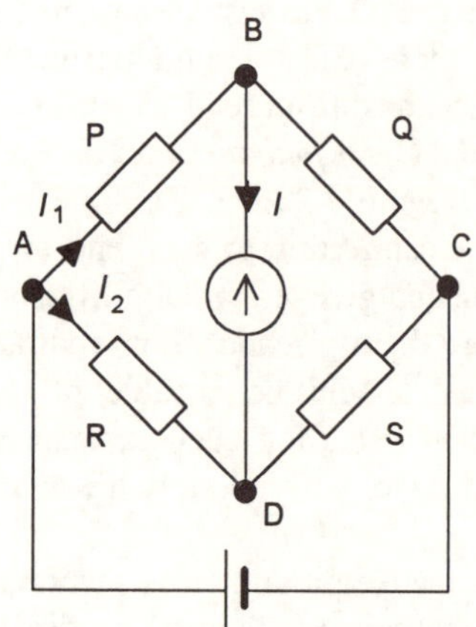

Figure 4.25 *Wheatstone bridge*

Activity

The *Wheatstone bridge* gives a more accurate method of measuring resistance than the ammeter–voltmeter method. Figure 4.25 shows the basic form of such a bridge. The resistances P, Q, R and S in the arms of the bridge are adjusted so that there is no current through the galvanometer. In such a condition the bridge is said to be balanced. In order to develop the theory of the bridge we have to consider the potential divider circuit shown in Figure 4.26(a). Because the two resistors are in series, the current through each of them will be the same. Thus the potential difference across R is I_2R and that across S is I_2S. The supply voltage V_s is thus divided between the two resistors according to the value of their resistances. Likewise, for the circuit shown in Figure 4.26(b), the current through P and Q will be the same and so the potential difference across P is I_1P and that across Q is I_1Q. The supply voltage V_s is thus divided between the two resistors according to their resistances. If we combine these two circuits, we have the Wheatstone bridge circuit.

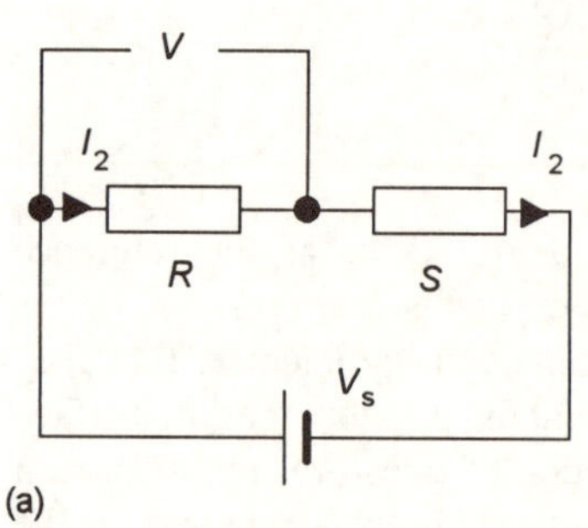

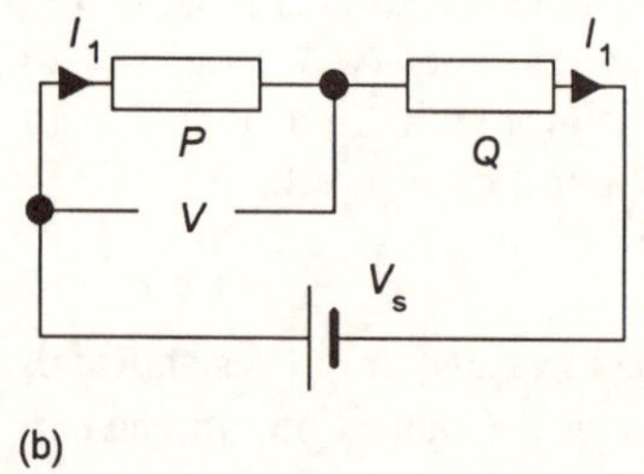

Figure 4.26 *Potential divider circuits*

When there is no current through the galvanometer (Figure 4.25), then there must be no potential difference across it. Points B and D must be at the same potential. Thus the potential difference across P must be the same as that across R, i.e. $I_1P = I_2R$, and so:

$$\frac{P}{R} = \frac{I_2}{I_1}$$

It also means that the potential difference across Q is the same as that across S. Thus, since there is no current through the galvanometer, the current through P must be the same as that through Q and the current through R the same as that through S. Hence $I_1 Q = I_2 S$ and so:

$$\frac{Q}{S} = \frac{I_2}{I_1}$$

Thus we must have:

$$\frac{P}{R} = \frac{Q}{S}$$

This balance condition is independent of the supply voltage, depending only on the resistances in the four arms of the bridge. The galvanometer only has to determine whether there is a current and so the result is not affected by the calibration of the instrument. It is termed a *null method* since zero current is being looked for. The Wheatstone bridge is thus capable of high precision and is widely used for the measurement of resistances in the range 1 Ω to 1 MΩ.

Figure 4.27 shows a version of the Wheatstone bridge called the *metre bridge*, use such a bridge to determine the resistance of a length of wire and hence its resistivity. The length of wire is P and an accurately known resistance is used for Q, e.g. a resistance box. The resistances R and S are provided by a length of uniform resistance wire of length L, typically one metre. The resistance per unit length r is thus the same at all points along the wire. The jockey, a knife-edged movable contact, is moved to a point along the wire that results in zero current through the galvanometer. If this occurs with the jockey a distance x from one end, then we have $R = xr$ and $S = (L - x)r$. Hence:

$$P = \frac{R}{S} \times Q = \frac{xr}{(L-x)r} \times Q = \frac{x}{L-x} \times Q$$

The greatest accuracy is obtained when the balance point is near the centre of L, both x and $(L - x)$ can then be determined to a reasonable degree of accuracy.

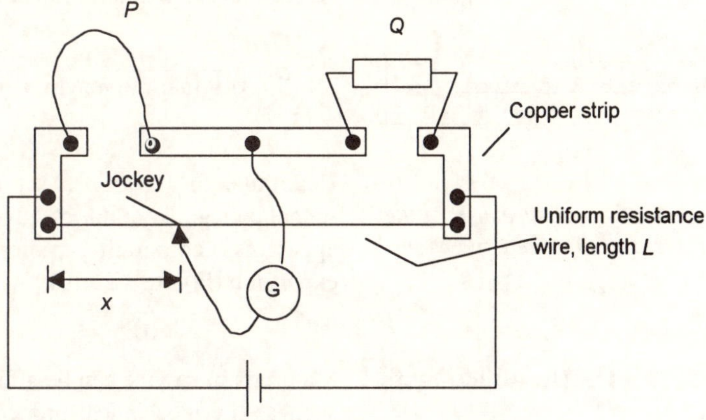

Figure 4.27 *The metre bridge*

Example

A 20 cm length of a copper–manganese alloy wire is placed in one gap of a metre bridge and a standard resistance of 10 Ω in the other gap. Balance is achieved when the jockey is 41.1 cm from the end of the wire, length 100.0 cm, nearest the unknown resistance end. Determine the resistance of the wire.

Using the equation developed above:

$$R_1 = \frac{x}{L-x} \times R_2 = \frac{41.1}{100.0-41.1} \times 10 = 6.98$$

4.7.1 Dielectric strength

The dielectric strength is the voltage needed per unit thickness of the material for electric breakdown. This can be measured by placing a sheet of the material between two conductors and increasing the voltage between the two until there is an increase in the current from a barely measurable value to quite a significant current. The increase in current occurs because at breakdown the material changes from being a very good insulator to a quite reasonable conductor.

4.8 Thermal expansion

The determination of the coefficient of linear expansion of a metal requires some method of measuring small changes in length. A basic arrangement is shown in Figure 4.28(a); a rod of length about 50 to 100 cm fits inside a steam jacket with one end of the rod against a rigid stop and the other end free to move. Near the free end there is a scratch on the rod; the length of the rod from the fixed end to the scratch is measured. A vernier microscope is focused on the scratch when the rod is at room temperature, the temperature being measured, and the vernier reading taken. Steam is then admitted to the chamber round the rod and, after long enough to ensure that the rod is at the steam temperature and the rod is no longer expanding, the temperature and vernier reading are taken. The coefficient of linear expansion a is then:

$$a = \frac{\text{change in vernier reading}}{\text{original rod length} \times \text{change in temp.}}$$

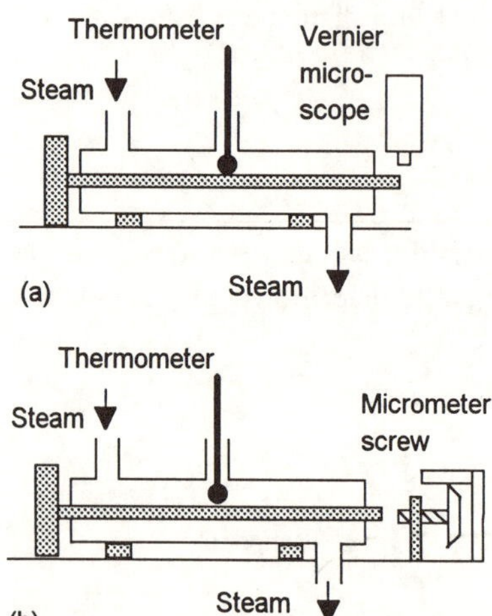

Figure 4.28 *Measurement of the coefficient of thermal expansion*

Activity

Determine the coefficient of linear expansion of a metal rod. Many school/college laboratories have permanently assembled apparatus, such apparatus incorporating a micrometer screw for the measurement of the expansion (Figure 4.28(b)).

4.9 Durability tests

Materials in service can be affected by their environment and properties, such as the mechanical properties of strength and toughness, changed. Measurements of these other properties thus give an indication of the interaction of the environment with a material.

Metallic materials corrode in moist air with some metals corroding at a faster rate than others. Corrosion testing is not a simple operation with field trials or such conditions simulated in the laboratory. Essentially the tests are the observation of what happens to the metals over a period of time. Metals exposed to corrosive environments are often protected by being coated with a material such as paint. Tests are then used to investigate the weathering characteristics of the painted material. An accelerated weathering process is often used with exposure to radiation from an electric arc, with intermittent exposure to a spray of water to simulate rain. BS 6917 gives details of corrosion testing in artificial atmospheres, indicating the requirements for specimens, apparatus and procedures. BS 3900: Part G gives details of environmental tests on paint films.

When metals are used at high temperatures is often restricted by surface attack or scaling which gradually reduces the cross-sectional area and hence the stress-bearing ability of the item. The build up of oxide layers at high temperatures is very much influenced by the environment, e.g. metal pipes exposed to superheated steam or hot gases from furnaces. Materials are tested by exposing them to such situations and measuring the reduction in the metal thickness as a consequence of the corrosive attack.

Plastic materials may dissolve in some liquids or absorb sufficient of the liquid to have their properties changed. When absorption occurs the plastic becomes permeable to the liquid, i.e. liquid can leak through it. This permeability is of vital concern if the plastic is being considered for used as a container for liquids, e.g. a Coca-Cola bottle (see Section 2.1).

Plastics are not generally subject to corrosion in the same way as metals but they can be adversely affected by weathering, i.e. exposure to light, heat, rain, sun. This can show itself as a fading of the colour of the plastic and/or a loss of flexibility. Tests are used to determine the weathering resistance of plastics with them being subject to an accelerated weathering process involving exposure to radiation from an electric arc, with intermittent exposure to a spray of water to simulate rain. The tests tend to be comparative ones with standard colours/materials being simultaneously exposed and performances compared.

Example

Tables indicate that a weight loss of 1 mg per exposed area of 0.01 m^2 (1 dm^2) per day for cast iron is a penetration of corrosion into the cast iron surface of 4.65 μm per year. What will be the penetration of corrosion into a cast iron product in a monitored situation if it suffers a weight loss of 0.5 mg/dm^2 in a day?

Since 1 mg/dm^2 in a day is 4.65 μm per year then 0.5 mg/dm^2 is 2.325 μm per year.

Example

The following data are test results on the corrosion rate for different metals suspended in the hot fumes from the combustion of fuel oils. On the basis of that data, coupled with the additional information supplied discuss the possible choice of a metal for pipes which would be exposed to the fumes.

	Corrosion rate in mm/year
Steel with 25% Cr, 20% Ni	Completely corroded
35% Cr-65% Ni alloy	Completely corroded
50% Cr-50% Ni alloy	4
60% Cr-40% Ni alloy	2

The 50% Cr-50% Ni alloy has a tensile strength of 550 MPa, a yield stress of 340 MPa and a Charpy impact strength of 37 J. The 60% Cr-40% Ni alloy has a tensile strength of 760 MPa, a yield stress of 590 MPa and a Charpy impact strength of 7 J.

On the basis of the corrosion tests the choice is between the 50% Cr-65% Ni and the 60% Cr-40% Ni alloys, with the latter having better corrosion properties. The mechanical properties indicate that the 60% Cr-40% Ni alloy is stronger but considerably more brittle. The more ductile and shock resistant properties are likely to mean that the 50% Cr-50 % Ni alloy is the choice.

Activity
Design a test that could be used for the measurement of the weathering of painted wood test pieces in order to test the effectiveness of different forms of paint and undercoating.

Activity
Design a test that could be used to test the durability of car bodywork when subject to the conditions cars are likely to encounter.

Problems

1 The following results were obtained from a tensile test of an aluminium alloy. The test piece had a diameter of 11.28 mm and a gauge length of 56 mm. Plot the stress-strain graph and determine (a) the tensile modulus, (b) the 0.1% proof stress.

Load/kN	0	2.5	5.0	7.5	10.0	12.5	15.0	17.5
Ext./mm	0	1.8	4.0	6.2	8.4	10.0	12.5	14.6

Load/kN	20.0	22.5	25.0	27.5	30.0	32.5	35.0
Ext./mm	16.3	19.0	21.2	23.5	25.7	28.1	31.5

Load/kN	37.5	38.5	39.0	39.0 (broke)
Ext./mm	35.0	40.0	61.0	86.0

2 The following results were obtained from a tensile test of a polymer. The test piece had a width of 20 mm, a thickness of 3 mm and a

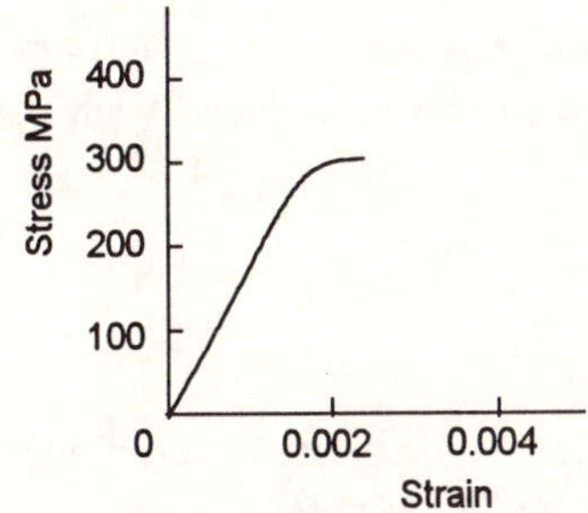

Figure 4.29 *Problem 6*

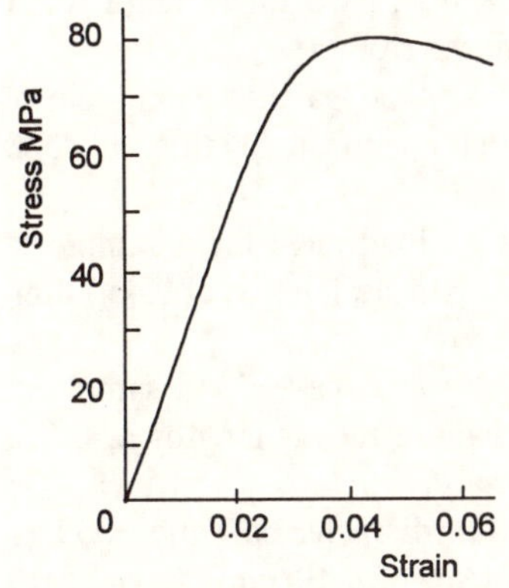

Figure 4.30 *Problem 7*

gauge length of 80 mm. Plot the stress-strain graph and determine (a) the tensile strength, (b) the secant modulus at 0.2% strain.

Load/N	0	100	200	300	400	500	600	630	650
Ext./mm	0	0.08	0.17	0.35	0.59	0.88	1.33	2.00	2.40

3 The following results were obtained from a tensile test of a steel. The test piece had a diameter of 10 mm and a gauge length of 50 mm. Plot the stress-strain graph and determine (a) the tensile strength, (b) the yield stress, (c) the tensile modulus.

Load/kN	0	5	10	15	20	25	30	32.5	35.8
Ext./mm	0	0.016	0.033	0.049	0.065	0.081	0.097	0.106	0.250

4 A flat tensile test piece of steel has a gauge length of 100.0 mm. After fracture, the gauge length was 131.1 mm. What is the percentage elongation?

5 The following data was obtained from a tensile test on a stainless steel test piece. Determine (a) the limit of proportionality stress, (b) the tensile modulus.

Stress/MPa	0	90	170	255	345	495	605
Strain/$\times 10^{-4}$	0	5	10	15	20	30	40

Stress/MPa	700	760	805	845	880	895
Strain/$\times 10^{-4}$	50	60	70	80	90	100

6 Estimate from the stress-strain graph for cast iron in Figure 4.29 the tensile strength and the limit of proportionality.

7 Estimate from the stress-strain graph for a sample of nylon 6 given in Figure 4.30 the tensile modulus and the tensile strength.

8 Sketch the form of the stress-strain graphs for (a) brittle stiff materials, (b) brittle non-stiff materials, (c) ductile stiff materials, (d) ductile non-stiff materials.

9 The effect of working an aluminium alloy (1.25% Mn) is to change the tensile strength from 110 MPa to 180 MPa and the percentage elongation from 30% to 3%. What is the effect of the working on the properties of the material?

10 An annealed titanium alloy has a tensile strength of 880 MPa and a percentage elongation of 16%. An annealed nickel alloy has a tensile strength of 700 MPa and a percentage elongation of 35%. Which alloy is (a) the stronger, (b) the more ductile in the annealed condition?

11 Cellulose acetate has a tensile modulus of 1.5 GPa and polythene a modulus of 0.6 GPa. Which of the two plastics will be the stiffer?

12 The following are Izod impact energies at different temperatures for samples of annealed cartridge brass (70% Cu-30% Zn). What can be deduced from the results?

Temperature °C	+27	−78	−197
Impact energy J	88	92	108

13 The following are Charpy V-notch impact energies for annealed titanium at different temperatures. What can be deduced from the results?

Temperature °C	+27	−78	−196
Impact energy J	24	19	15

14 The following are Charpy impact strengths for nylon 6.6 at different temperatures. What can be deduced from the results?

Temperature °C	−23	−33	−43	−63
Impact strength kJ/m^2	24	13	11	8

15 The impact strengths of samples of nylon 6, at a temperature of 22°C, are found to be 3 kJ/m^2 in the as-moulded condition but 25 kJ/m^2 when the sample has gained 2.5% in weight through water absorption. What can be deduced from the results?

16 With the Vickers hardness test a 30 kg load gave for a sample of steel an indentation with diagonals having mean lengths of 0.530 mm. What is the hardness?

17 With the Vickers hardness test a 30 kg load gave for a sample of steel an indention with diagonals having mean lengths of 0.450 mm. What is the hardness?

18 With the Vickers hardness test a 10 kg load gave for a sample of brass an indentation with diagonals having means lengths of 0.510 mm. What is the hardness?

19 With the Brinell hardness test a 10 mm diameter ball and 3000 kg load gave an indentation with a diameter of 4.10 mm. What is the hardness?

20 With the Brinell hardness test a sample of cold worked copper with a 1 mm diameter ball and 20 kg load gave an indentation of diameter 0.630 mm. What is the hardness?

21 The dielectric strength of a plastic was measured as 31 kV/mm when dry and after 2 days exposed to 80% humidity as 29 kV/mm. Explain the significance of the data?

22 After 4000 hours exposure to the fumes in an oil-fired furnace samples of metals were found to show the following corrosion rare. Explain the significance of the data?

Steel 25%Cr-12%Ni	0.11 mm/year
Steel 25%Cr-20%Ni	0.28 mm/year
65%Ni-35%Cr alloy	0.02 mm/year

23 Oxide penetration on steels exposed for 5000 hours to steam at about 600°C was found to be as follows. Discuss the significance of the data.

0.11%C steel	0.40 mm
0.34%C steel	0.25 mm
1.24%Cr-0.5%Mo-1.4%Si steel	0.12 mm
2.25%Cr-0.5%Mo-0.75%Si steel	0.09 mm

24 The corrosion rate for mild steel test plates was found to give averages of 0.050 mm per year in rural surroundings, 0.070 mm per year in marine surroundings and 0.150 mm per year in heavy industrial surroundings. Discuss the significance of the data.

25 Specify the type of test that can be used in the following instances:

(a) A storekeeper has mixed up two batches of steel, one batch having been surface hardened and the other not. How could the two be distinguished?

(b) What test could be used to check whether tempering has been correctly carried out for a steel?

(c) A plastic is modified by the inclusion of glass fibres. What test can be used to determine whether this has made the plastic stiffer?

(d) What test could be used to determine whether a metal has been correctly heat treated?

(e) What test could be used to determine whether a metal is in a suitable condition for forming by bending?

5 Structure and properties of solids

5.1 The crystalline structure

Why are polymers weaker and less stiff than metals? Why are ceramics so stiff? How can we make a metal more ductile? In order to answer these and other questions, and gain an understanding of the properties of materials, we need to consider the structure of materials.

Materials which have their particles arranged in a regular repetitive arrangement which extends throughout the material are termed *crystalline*; materials where there is no such orderly arrangement are said to be *amorphous*.

5.1.1 Orderly packing of spheres

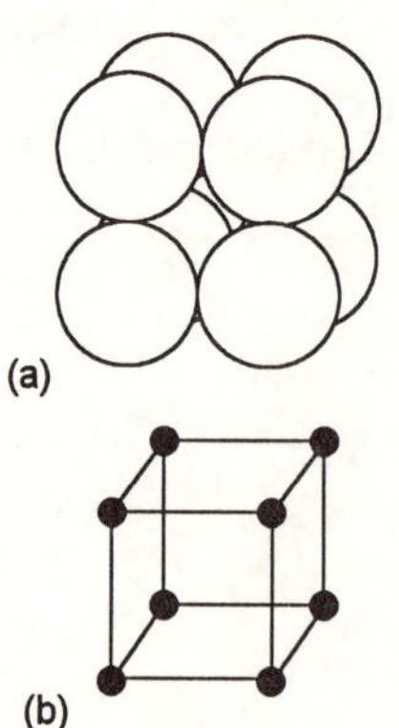

Figure 5.1 *Simple cubic structure*

Consider the stacking together of spheres in an orderly manner. Since spheres can be stacked in any way, a sphere can be considered to be a model for an atom, ion or molecule in a solid when there is no directionality to the bonding forces. One of the simplest arrangement of spheres is that of the *simple cubic structure*. Figure 5.1(a) shows the structure obtained by stacking four spheres with the centres of each sphere at the corners of a cube. The surfaces of each sphere touch the surfaces of each of its neighbours in such a way that the length of the side of the cube is equal to the diameter of the spheres. The line joining the centres of the spheres (Figure 5.1(b)) encloses what is termed the *unit cell*, this being the smallest arrangement of particles that when regularly repeated forms the crystal. The resulting solid would consist of a completely orderly array of spheres, i.e. particles, and we would expect the surfaces of such a solid to be smooth and flat with the angles between adjoining faces always 90°. Such a solid would when broken up always have the appearance of stacked cubes. This is a description of a *cubic crystal*.

The simple cubic crystal shape is arrived at by stacking spheres in one particular way. By stacking spheres in a closer manner other crystal shapes can be produced: body-centred cubic, face-centred cubic and hexagonal close-packed structures. These three close-packed structures represent the structures occurring with solid metals.

With the *body-centred cubic* unit cell (Figure 5.2), a square arrangement of spheres is used in each layer but successive layers are displaced so that spheres fit within the hollows of the layer underneath. As a consequence, the arrangement is slightly more complex than the simple cubic unit cell in having an extra sphere in the centre of the cell.

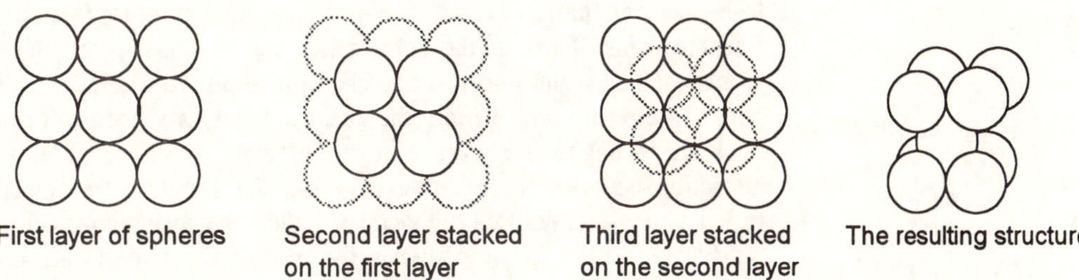

Figure 5.2 *Body-centred cubic structure*

With the *face-centred cubic* unit cell (Figure 5.3), the spheres in each layer are packed as close as possible and successive layers are displaced so that spheres fit within the hollows of the layer underneath. There is, when compared with the simple cubic unit cell, a sphere at the centre of each face of the cube. With the *hexagonal close-packed* unit cell (Figure 5.4), the spheres are packed in a close array which gives a hexagonal form of structure.

An important point to notice with all these structures is that there are spaces between the spheres in the crystal structures. The size of these spaces depends on the type of structure. Within these spaces it is possible to fit other atoms without too much strain on the crystalline structure.

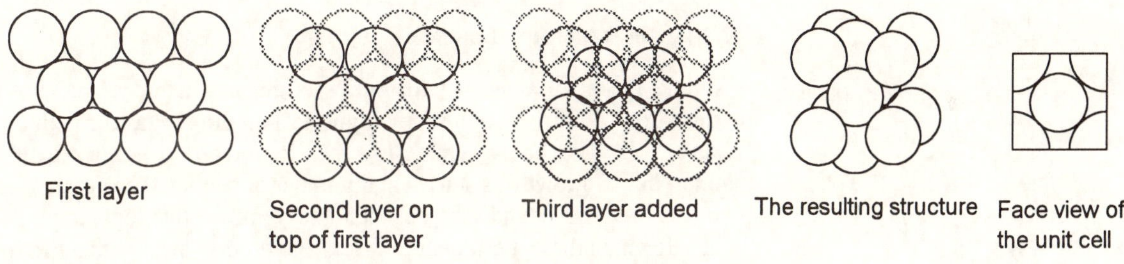

Figure 5.3 *Face-centred cubic structure*

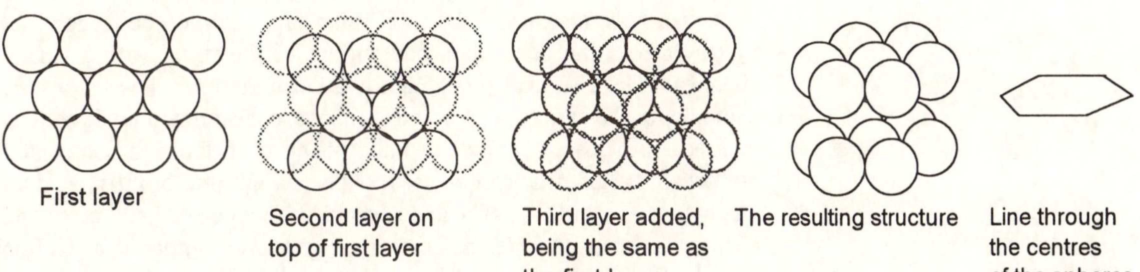

Figure 5.4 *Hexagonal close-packed structure*

5.2 Structure of metals

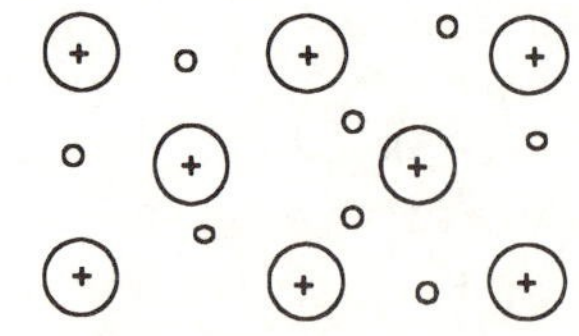

Figure 5.5 *Metals*

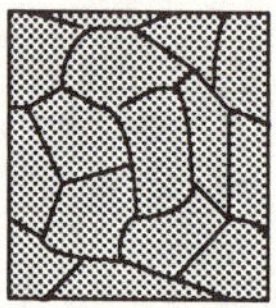

Figure 5.6 *Grains*

Metals are elements, such as copper, which have atoms which so readily lose electrons that, in the solid state at room temperature, there are many free electrons. Thus in the solid state, copper consists of an array of atoms each of which has lost one electron (Figure 5.5). This leaves each copper atom as a positive ion; the electrons that have been lost remain as a cloud if negative charge floating between the ions. The force of attraction between the positive ions and the negative free electrons is rather like a glue holding the metal together. Because the resulting bonds holding the metal ions together in the solid is the same in all directions round each ion, a simple model for a metal ion is a sphere. The free electrons explain why metals are good conductors of electricity, since they have free charge carriers which are easily moved by the application of a voltage.

Metals are crystalline substances. This may seem a strange statement in that metals do not generally look like crystals, with their geometrically regular shapes. However, if we consider a metal in solidifying from the liquid as not growing as a single crystal but having crystals starting to grow at a number of points within the liquid, then the result is a mass of crystals. Each crystal is prevented from reaching geometrically regular shapes by neighbouring crystals growing and restricting its growth (Figure 5.6). The result is said to be *polycrystalline* and the term *grain* is used to describe regions in the metal for which there are orderly arrangements of particles; at the boundaries between the grains the regular pattern breaks down as the pattern changes from the orderly arrangement of one grain to that of the next grain.

5.2.1 Grain structure of metals

A simple model of a metal with grains is given if a raft of bubbles is produced on the surface of a soapy liquid by bubbling a gas through a jet (Figure 5.7). The bubbles pack together in an orderly and repetitive manner, but if 'growth' is started at a number of centres then 'grains' are produced. At the boundaries between the 'grains' the regular pattern breaks down as the pattern changes from the orderly arrangement in one 'grain' to the orderly arrangement in the next 'grain' and gives a good representation of a metal.

The grains in the surface of a metal are not generally visible, though an exception is the very large grains which are readily visible in the surface of galvanised steel objects. Grains can, however, be made visible by careful etching of the polished surface of the metal with a suitable chemical. The chemical preferentially attacks the grain boundaries. For example, in the case of copper and its alloys, concentrated nitric acid can be used. In the case of carbon and alloy steels of medium carbon content, an etchant called nital can be used. Nital is a mixture of nitric acid and alcohol, typically 5 ml of acid to 95 ml of alcohol. Details of suitable chemicals are given in reference books, e.g. *Metals Databook* by C. Robb (The Institute of Metals 1987); proper safety precautions in the handling and disposing of the chemicals are vital since they are highly corrosive and many of them are potentially lethal.

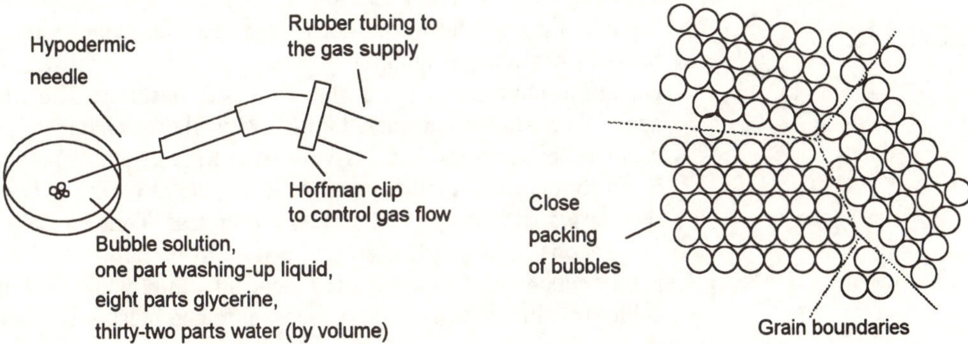

Figure 5.7 *(a) Simple arrangement for producing bubbles, (b) a raft of bubbles*

5.3 Stretching metals

A simple way we can think of the atoms in a metal is as though there were an array of spheres tethered to each other by springs, as illustrated in Figure 5.8(a). When forces are applied to stretch the material then the springs are stretched and exert an attractive force pulling the atoms back to their original positions. When forces are applied to compress the material then the springs are compressed and exert repulsive forces which push the atoms back to their original positions (Figure 5.8(b)). We thus have a simple model to explain the elastic properties of metals.

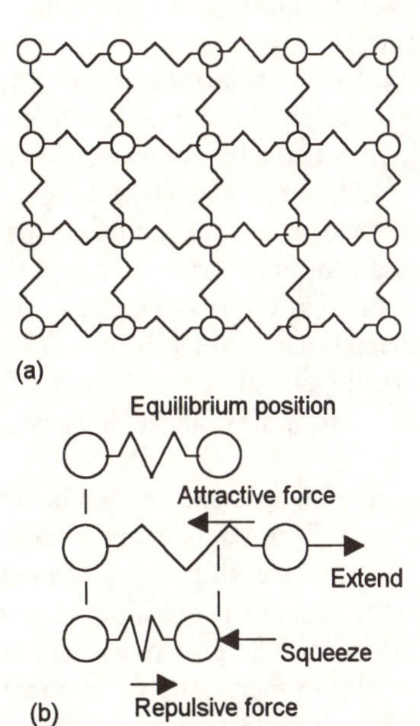

Figure 5.8 *A model of a metal*

5.3.1 Elastic and plastic behaviour

A simple theory to explain the elastic and plastic behaviour of metals is the *block slip model*. A metal is considered to be made of blocks of atoms which can be made to move relative to each other by the application of shear stress (Figure 5.9). Thus plastic deformation, i.e. slip of atoms to new permanent positions, occurs in a single movement involving one whole plane of atoms sliding bodily over another. With this model we can only have slip within an orderly arrangement of atoms, i.e. within a grain. Slip planes cannot thus cross over from one grain to another; the disorderly arrangement at the grain boundaries does not allow it. Thus a metal having large grains can give more slippage, i.e. permanent deformation, and so be more ductile, than one with small grains.

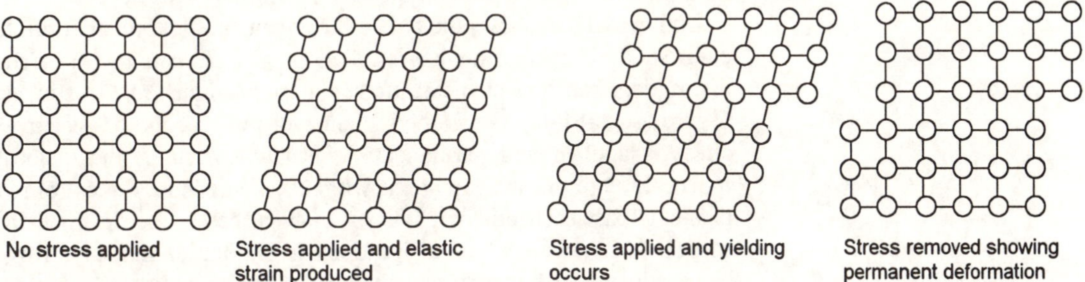

Figure 5.9 *Block slip model*

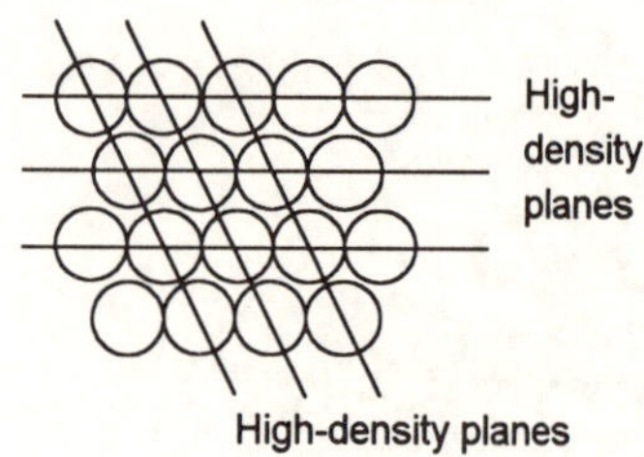

Figure 5.10 *High-density planes*

While there can be considered to be many planes of atoms in a crystal, slip is found to only occur between the planes with the closest packing of atoms (Figure 5.10). This is because the atoms are close enough to more easily permit of changes of positions than when further apart. The number of such high-density planes along which slip can occur depends on the form of structure of the crystal. The body-centred cubic structure has many such slip planes, the face-centred cubic less and the hexagonal close-packed structure even less. Thus metals which have a hexagonal close-packed structure tend to be harder and more brittle than metals with the face centred cubic structure, while the body-centred cubic structure is likely to be the least hard and most ductile metal.

5.3.2 Dislocation model

The above is just a simple model of what happens when metals are stretched. The model needs modification to do more than give a simple idea of what happens. The above model has assumed that the arrangement of atoms within a grain is perfectly orderly. In reality this is not the case and there are some atoms in the wrong places, the term *dislocations* being used. If, however, we consider the arrangement to be imperfect then permanent deformations can be produced with much less stress. When you have a large carpet which is perfectly flat on the floor, it requires quite an effort to slide the entire carpet and make it move across the floor (Figure 5.11(a)). But if there is a ruck in the carpet (Figure 5.11(b)), then the carpet can be slid over the floor by pushing the ruck along a bit at a time and considerably less effort is required. An entire plane is not being moved at one instant, just a little bit of it. This is the type of movement which is considered to take place within a metal, the 'ruck' in the lattice being a *dislocation* of atoms due to imperfect packing of the atoms within the lattice.

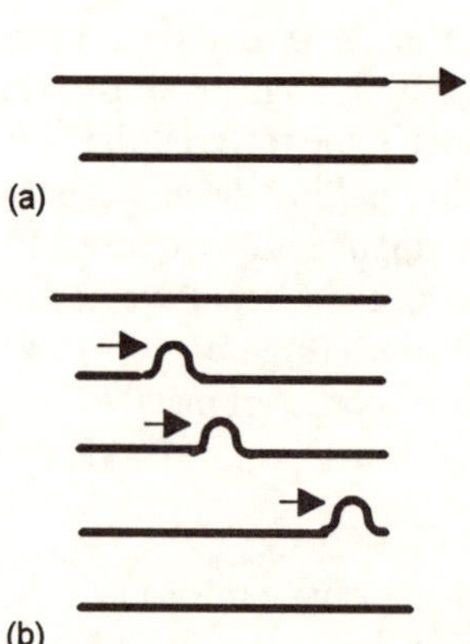

Figure 5.11 *Moving a carpet*

Dislocations are produced as a result of missing atoms, atoms being displaced from their correct positions, and foreign atoms being present and distorting the orderly packing of atoms. Cold working which distorts grains results in an increase in dislocations because it displaces atoms from their correct positions. The foreign atoms may be present as a result of a deliberate alloying process. Thus alloying in increasing the number of dislocations and making it more difficult for dislocations to move through the material increases the yield stress. Another method of increasing the yield stress is to cause small particles to precipitate out from an alloy. Such a process is called *precipitation hardening*.

Figure 5.12(a) shows the type of arrangement of atoms that might be considered to occur with what is called an *edge dislocation*. With such a dislocation, the line of dislocation is at right angles to the slip plane. Figure 5.12(a)(b)(c) shows the sequence of movements that occur when stress is applied and permanent deformation occurs. The dislocation moves through the array of atoms without wholesale movement of planes of atoms past each other; it is a bit-by-bit process like the ruck in the carpet. The movement of dislocations is similar to that given by the block slip model but requires smaller stresses as only a few bonds are being altered or broken at any one time.

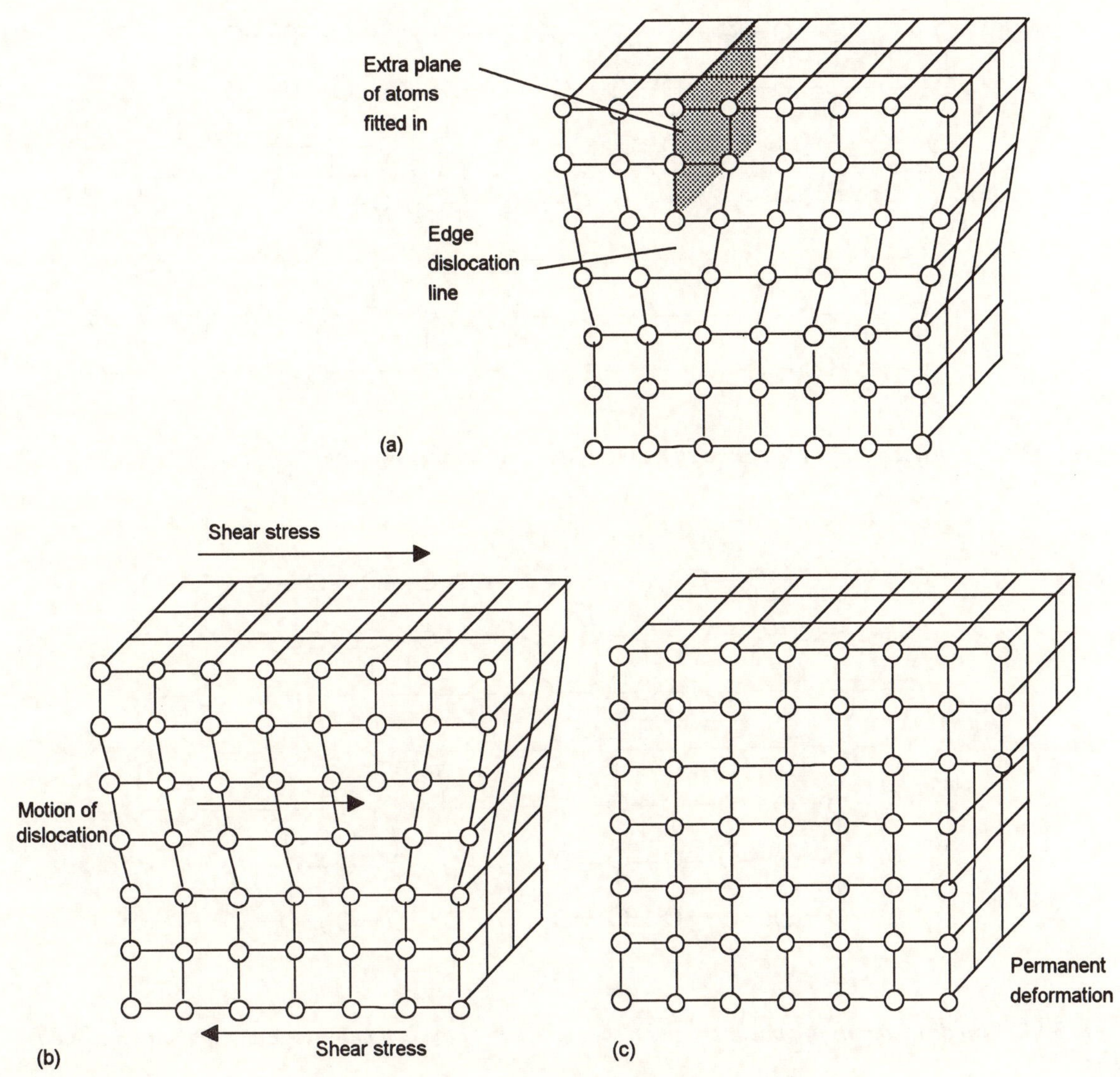

Figure 5.12 *Shear stress causing deformation when an edge location is present*

What happens when two dislocations come close to each other during their movement through a metal? With a dislocation, the atoms on one side of the slip plane are in compression and on the other side in tension. When two dislocations come together, as in Figure 5.13, the regions of compression can impinge on each other and so hinder the movement of the dislocations. If the movement of the dislocations is such as to bring the compression region of one dislocation against the tension region of another dislocation (Figure 5.14) then it is possible for the two dislocations to annihilate each other.

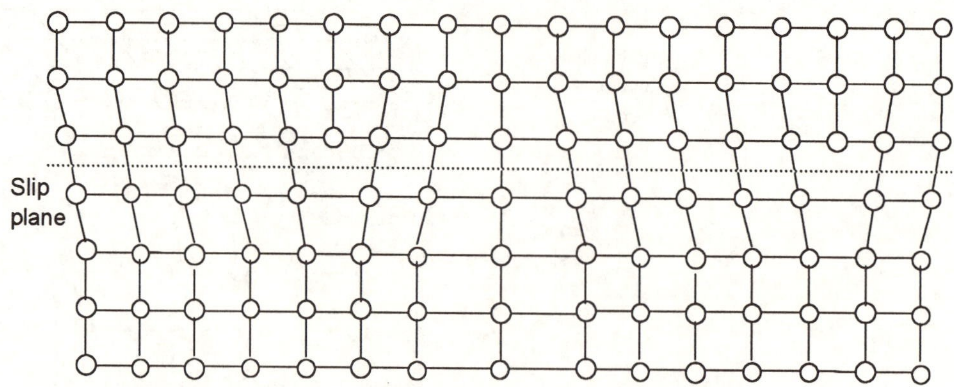

Figure 5.13 *Two dislocations of the same sign on the same slip plane 'repelling' each other*

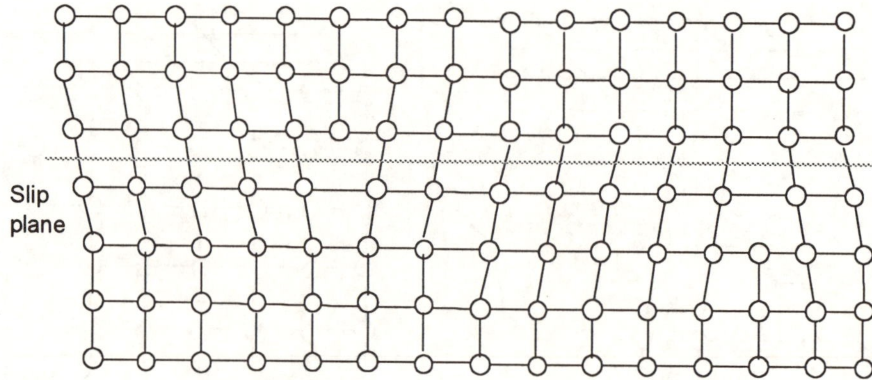

Figure 5.14 *Two dislocations of opposite sign on the same slip plane can move together and annihilate each other*

In general, the more dislocations a metal has, the more the dislocations get in the way of each other and so the more difficult it is for the dislocations to move through the metal. More stress is needed to cause yielding. *Work hardening* occurs as a result of a material being plastically deformed, this increasing the dislocation density.

The movement of dislocations through a metal is also hindered by the grain boundaries. The more grain boundaries there are in a metal, the more difficult it is to produce yielding. More grain boundaries occur when the grain size in a metal is small, thus a treatment which reduces grain size makes a metal stronger while one which increases grain size makes it weaker.

The movement of dislocations is hindered by anything that destroys the continuity of the atomic array. The presence of 'foreign' atoms can distort the atomic array of a metal and so hinder the movement of dislocations. *Dispersion hardening* increases the yield stress of a metal by producing a dispersion of fine particles throughout the material, these hindering the movement of dislocations. *Alloying* involves the introduction of foreign atoms into a crystal lattice, producing interstitial and substitutional point defects which hinder the movement of dislocations. Hence alloys tend to have a higher yield stress than the parent metal alone, indeed pure metals like copper and iron are very soft and no use as an engineering material but when alloyed become much stronger. This is referred to as *solution hardening*.

5.4 Alloys

One way of modifying the properties of a metal is alloying. An *alloy* is a metallic material made by mixing two or more elements. The everyday metallic objects around you will be made, almost invariably, from alloys rather then the pure metals themselves. Pure metals do not always have the appropriate combinations of properties needed; alloys can, however, be designed to have them.

Making alloys is rather like baking a cake. The basic ingredients of flour, sugar, fat, eggs and water are mixed together and then cooked. The result is a cake which has a texture and properties quite different from those of the individual ingredients. The type of cake produced depends on the relative amounts of the ingredients and the way it is cooked. In making alloys, the ingredients are mixed and heated and the resulting alloy can have properties quite different from those of the ingredients. The properties will depend on the relative amounts and nature of the ingredients as well as how they are 'baked'. An alloy is a particular mixture of components and so has a particular chemical composition, e.g. one carbon steel may be 99.0% iron combined with 1.0% carbon while another is 99.5% iron with 0.5% carbon.

The coins in your pocket are made of alloys. Coins need to be made of a relatively hard material which does not wear away rapidly, i.e. the coins have to have a 'life' of many years. Coins made of, say, pure copper would be very soft; not only would they suffer considerable wear but they would bend in your pocket. The copper-looking British coins are made of an alloy of copper with 2.5% by weight of zinc and 0.5% of tin, the term coinage bronze being used for the alloy. The silver-looking British coins are made of an alloy of copper with 25% by weight of nickel, the term cupro-nickel being used for the alloy.

Pure metals tend to be soft with high ductility, low tensile strength and low yield strength. Because of this they are rarely used in engineering. Alloying can produce harder materials with higher tensile strength, higher yield stress and a reduction in ductility. Such materials are more useful in engineering. There are, however, some circumstances in which the properties of pure metals are useful. They are where high electrical conductivity is required (alloying reduces conductivity); where good corrosion resistance is required (alloying can result in less corrosion resistance); and where very high ductility is required.

We can think of the structure of alloys in terms of the constituent metals, say A and B, being mixed in the liquid state. Then when the

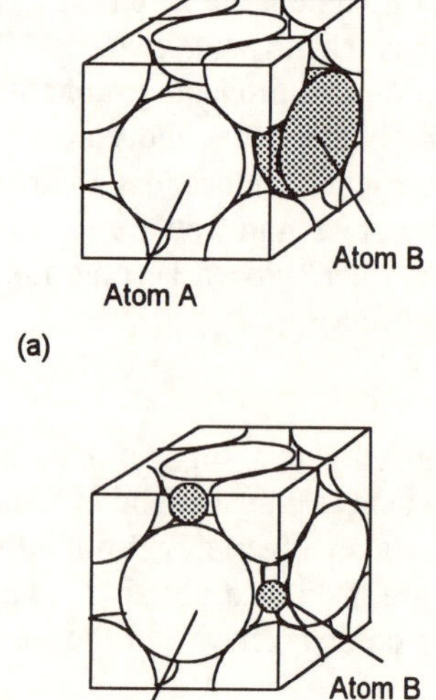

Figure 5.15 *Possible forms of alloys*

mixture solidifies, there is the possibility that solid alloy will have a crystal structure in which some of the atoms in the crystal structure of A have been replaced by atoms of B (Figure 5.15(a)). Alternatively, because there are spaces between the atoms of A in its crystal structure, some atoms of A, if small enough, might lodge in these spaces (Figure 5.15(b)). Another possibility is that elements A and B combine to form a chemical compound. With a compound there will be a particular structure for that compound with atoms of A and B assuming specific positions, rather than just popping into any gap. Another possibility is that when the liquid mixture cools A and B separate out, with B forming its own crystal structure independent of A. The structure then becomes a mixture of two types of crystals.

5.4.1 Ferrous alloys

Pure iron is a relatively soft material and is hardly ever used. Alloys of iron with carbon are, however, very widely used, the term *ferrous alloys* being used for alloys with iron. Pure iron at room temperature exists as a body-centred cubic structure, this being commonly referred to as *ferrite*. This form continues to exist up to 912°C. At this temperature the structure changes to a face-centred cubic one, known as *austenite*. Iron atoms have a diameter of 0.256 nm (1 nanometre = 10^{-9} m); carbon atoms are much smaller with a diameter of 0.154 nm. The face-centred cubic structure is a more open structure than the body-centred structure; the face-centred structure of austenite has voids which can accommodate spheres up to 0.104 nm in diameter while the body-centred cubic structure voids between the atoms which are 0.070 nm in diameter. Thus, carbon atoms can be more easily accommodated within austenite, without severe distortion of the lattice, than ferrite. Austenite can take up to 2.0% of carbon while ferrite can only take 0.2%. Thus when iron containing carbon is cooled from the austenite state to the ferrite state, there is a reduction in the amount of carbon that can be accommodated within the iron and so some of the carbon atoms come out of the crystals and form a compound, another crystal structure, between iron and carbon called *cementite*. Cementite is hard and brittle. The result can be a structure consisting of purely ferrite grains mixed with grains which have a laminated structure of ferrite and cementite. Such a laminated structure is termed *pearlite*. Pure cementite is harder than pearlite, which in turns is harder than pure ferrite. Thus the structure, and hence the properties, of the iron alloy is determined by the amount of carbon present.

The percentage of carbon alloyed with iron has a profound effect on the properties of the alloy. The terms used for the alloys produced with different percentages of carbon are:

Wrought iron	0 to 0.05% carbon
Steel	0.05 to 2% carbon
Cast iron	2 to 4.5% carbon

The term *carbon steel* is used for those steels in which essentially just iron and carbon are present. The term *alloy steel* is used where other elements are included.

5.4.2 Plain-carbon steels

Carbon steels are grouped according to their carbon content with the designations being roughly:

Mild steel	0.10 to 0.25% carbon
Medium-carbon steel	0.20% to 0.50% carbon
High-carbon steel	More than 0.50% carbon

Mild steel has a structure consisting predominantly of ferrite, medium-carbon steels tend to have about equal amounts of ferrite and pearlite, while high-carbon steels have predominantly pearlite with some free cementite occurring at high carbon contents.

Figure 5.16 shows how the mechanical properties of carbon steels depend on the percentage of carbon.

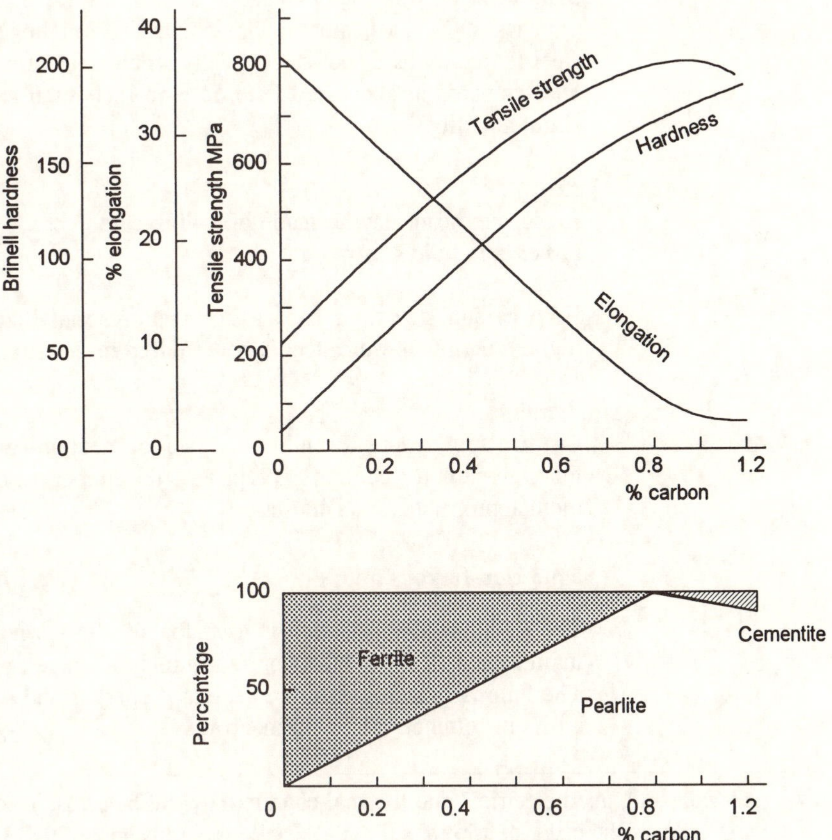

Figure 5.16 *Properties of carbon steels*

Increasing the percentage of carbon, within the range considered, increases the amount of pearlite at the expense of the softer ferrite, and hence:

1. Increases the tensile strength
2. Increases the hardness
3. Decreases the percentage elongation
4. Decreases the impact strength

Mild steel is a general purpose steel and is used where hardness and tensile strength are not the most important requirements but ductility is often required. Typical applications are sections for use as joists in buildings, bodywork for cars and ships, screws, nails, wire. Medium-carbon steel is used for agricultural tools, fasteners, dynamo and motor shafts, crankshafts, connecting rods, gears. With such steels the lower ductility puts a limit on the types of processes that can be used. Medium-carbon steels are capable of being quenched and tempered to develop reasonable toughness with strength. High-carbon steel is used for withstanding wear, where hardness is a more necessary requirement than ductility. It is used for machine tools, saws, hammers, cold chisels, punches, axes, dies, taps, drills, razors. The main use of high-carbon steel is mainly as a tool steel. High-carbon steels are usually quenched and tempered at about 250°C to develop their high strength with some slight ductility.

Example
A pickaxe head may be made of a high-carbon steel. Why high carbon rather than mild steel?

High carbon steel is a harder, stronger material than mild steel. The higher ductility of mild steel is not required in this situation.

Activity
List two applications for mild steel, two for medium-carbon steel and two for high-carbon steel and explain why the properties of the steels make them appropriate for such uses.

5.4.3 Non-ferrous alloys

The term *non-ferrous alloy* is used for all alloys where iron is not the main constituent, e.g. alloys of aluminium, of copper, of magnesium, etc. The following are some of the general properties and uses of non-ferrous alloys in common use in engineering.

Aluminium alloys	Low density, good electrical and thermal conductivity, high corrosion resistance. Tensile strengths of the order of 150 to 400 MPa, tensile modulus about 70 GPa. Used for metal boxes, cooking utensils, aircraft bodywork and parts.

Copper alloys	Good electrical and thermal conductivity, high corrosion resistance. Tensile strengths of the order of 180 to 300 MPa, tensile modulus about 20 to 28 GPa. Used for pump and valve parts, coins, instrument parts, springs, screws.
Magnesium alloys	Low density, good electrical and thermal conductivity. Tensile strengths of the order of 250 MPa and tensile modulus about 40 GPa. Used as castings and forgings in the aircraft industry where weight is an important consideration.
Nickel alloys	Good electrical and thermal conductivity, high corrosion resistance, can be used at high temperatures. Tensile strengths between about 350 and 1400 MPa, tensile modulus about 220 GPa. Used for pipes and containers in the chemical industry where high resistance to corrosive atmospheres is required, food processing equipment, gas turbine parts.
Titanium alloys	Low density, high strength, high corrosion resistance, can be used at high temperatures. Tensile strengths of the order of 1000 MPa, tensile modulus about 110 GPa. Used in aircraft for compressor discs, blades and casings, in chemical plant where high resistance to corrosive atmospheres is required.
Zinc alloys	Low melting points, good electrical and thermal conductivities, high corrosion resistance. tensile strength about 300 MPa, tensile modulus about 100 GPa. Used for car door handles, toys, car carburettor bodies – components that in general are produced by pouring the liquid metal into dies.

As an example of a non-ferrous alloy, consider copper alloys. Pure copper is a soft material with low tensile strength. For many engineering purposes it is alloyed with other metals. The exception is where high electrical conductivity is required. Pure copper has a better conductivity than the alloys. The following indicates the names given to the various types of copper alloys:

Copper with zinc	Brasses
Copper with tin	Bronzes
Copper with tin and phosphorus	Phosphor bronzes
Copper with tin and zinc	Gunmetals
Copper with aluminium	Aluminium bronzes
Copper with nickel	Cupro-nickels
Copper with zinc and nickel	Nickel silvers
Copper and silicon	Silicon bronze
Copper and beryllium	Beryllium bronze

Figure 5.17 shows how the percentage of zinc included with brasses affects the mechanical properties. Brasses with between 5 and 20% zinc are called gilding metals and, as the name implies, are used for architectural and decorative items to give a 'gilded' or golden colour. Cartridge brass is copper with 30% zinc. One of its main uses is for cartridge cases, items which require high ductility for the deep drawing process used to make them. The term common or basis brass is used for copper with 37% zinc. This is a good alloy for general use with cold

working processes and is used for fasteners and electrical connectors. It does not have the high ductility of those brasses with less zinc. Copper with 40% zinc is called Muntz metal.

The changes in the properties of brasses when the amount of zinc is changed arises from changes in the structure. Brasses with between 0 and 35% zinc form one type of structure, termed alpha, 5% and 45% there is a mixture of this alpha structure and another structure termed beta. It is this change in structure, i.e. the way the atoms of copper and zinc are packed together, that is responsible for the abrupt changes in properties of brass at 35% zinc.

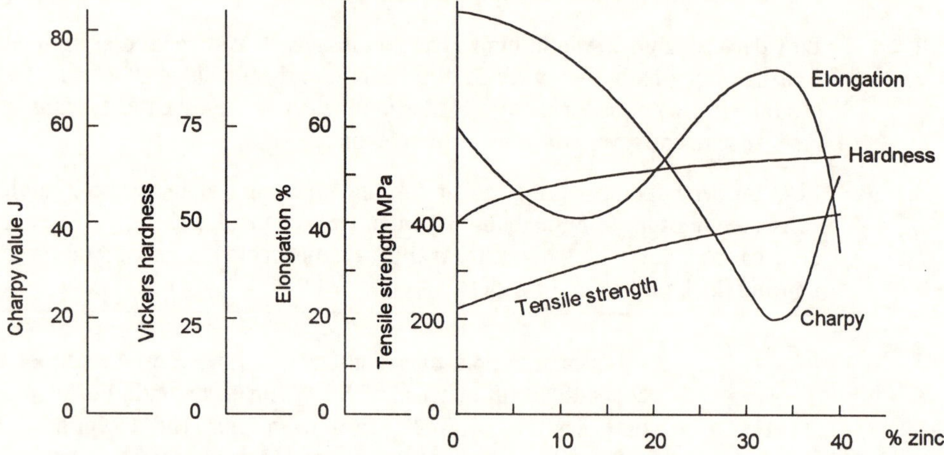

Figure 5.17 *Properties of brasses*

5.5 Heat treatment

One way in which the properties of a metal can be changed is by the use of *heat treatment*; heat treatment can be defined as the controlled heating and cooling of metals in the solid state for the purpose of altering their properties. A heat-treatment cycle consists normally of three parts:

1. Heating the metal to the required temperature for the changes in structure within the material to occur.
2. Holding at that temperature for a long enough time for the entire material to reach the required temperature and the structural changes to occur through the entire material.
3. Cooling, with the rate of cooling being controlled since it affects the structure and hence properties of the material.

5.5.1 Annealing

Annealing is the heat treatment used to make a metal softer and more ductile. It involves heating the metal to a high enough temperature for the grains to reform; this temperature is called the recrystallisation temperature and is about a third to half of the melting temperature of the

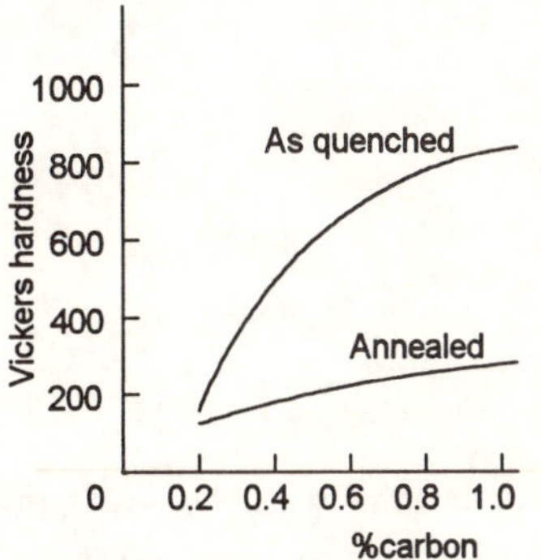

Figure 5.18 *Hardness of carbon steels*

metal. This heating is then followed by slow cooling. The result is a regrowth of grains to give a large grain structure.

In the case of carbon steels, this change in grain size is also accompanied by changes in the form of the constituents present in the alloy. Heating a steel to above the recrystallisation temperature changes the crystal structure from ferrite to austenite (see Section 5.4.1). The austenite can contain more carbon atoms than the ferrite. There is time with annealing, because the cooling is slow, for the excess carbon atoms to move out of the crystal structure to form a compound called cementite.

5.5.2 Quenching

If, after heating to a high enough temperature to cause recrystallisation, the heated carbon steel is cooled very rapidly, for example by being dropped into cold water, there is not time for the austenite to loose its excess carbon atoms and they become trapped in the structure. The result is a new structure called *martensite*. The carbon trapped in this structure considerably distorts the structure. As a consequence, martensite is very hard and brittle and the steel becomes harder and more brittle. The hardness and strength increases quite significantly with an increase in carbon content, this being because more carbon is trapped in the structure and so there is more distortion. This form of heat treatment is called *quenching*. Figure 5.18 shows how the hardness of carbon steels after quenching compares with that of the annealed steels.

A problem with the severe cooling that occurs in quenching is that cracking can occur. The cracks are a consequence of the stresses occurring as a result of the distortion produced by the structural changes and also differential expansion as a result of different parts of a product cooling at different rates.

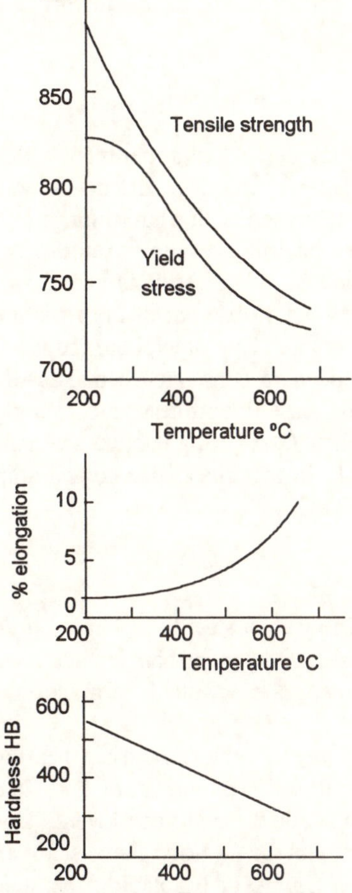

Figure 5.19 *The effect of tempering on the properties of a steel*

5.5.3 Tempering

In the quenched state carbon steels have such a low ductility as to be very difficult to use. The process known as *tempering* can, however, be used to improve the ductility without loosing all the hardness gained by the quenching.

Tempering involves heating the steel to a temperature at which some of the carbon trapped in the martensite structure can diffuse out and form cementite, so reducing the distortion of the structure. The amount of carbon that diffuses out depends on the temperature used for the tempering. Thus the mechanical properties depend on the tempering temperature. Figure 5.19 shows this for an alloy steel (a manganese-nickel-chromium-molybdenum steel).

Example

The following are the mechanical properties of a carbon steel. Identify the form of the internal structure which is responsible for these properties.

	Strength MPa	Hardness BH	Elongation %
As supplied	815	240	17
Annealed	625	180	23
Quenched and tempered at 200°C	1100	320	13
Quenched and tempered at 650°C	800	230	23

Annealing the material results in recrystallisation with the result that large grains are produced with a drop in the number of dislocations. As a result the material is weaker, softer and more ductile. Quenching results in the formation of martensite in which carbon atoms are trapped within a distorted structure. As a consequence the material is much harder and more brittle. Tempering allows some of the carbon atoms to diffuse out and reduce the distortion, the higher the tempering temperature the greater the reduction. Hence tempering restored some of the ductility but reduces the strength.

5.5.4 Precipitation hardening

A wide range of alloys used in engineering depend on a treatment called *precipitation hardening* for improvements in their hardness and strength. This type of treatment is widely used with aluminium alloys and nickel alloys. The process involves heating the alloy to above the recrystallisation temperature, then quenching. The result is a distorted crystal structure. However, with time, atoms diffuse out of the structure of this type of alloy to give a fine precipitate. This precipitate lodges at grain boundaries and in dislocations and as a consequence makes slip more difficult. The consequence is an increase in hardness and strength. For example, the aluminium-copper alloy (2014) might have a tensile strength of 185 MPa and hardness 45 HB in the annealed state and after precipitation hardening 425 MPa and 105 HB.

5.5.5 Surface hardening

There is often a need for the surface of a piece of steel to be hard, e.g. to make it wear resistant, without the entire component being made hard and often too brittle. Several methods are available for surface hardening.

For carbon steels surface hardening can be achieved by just heating the surface layers to above the recrystallisation temperature and then quenching to give a martensitic structure for these surface layers. This selective heating of the surface layers can be heating them with an oxyacetylene flame, so called *flame hardening*, or by placing the steel component in a coil carrying a high frequency current and allowing the induced currents in the surface layers to do the heating, so called *induction heating*. Another method that can be used to produce martensite in the surface layers is to increase the carbon content of the surface layers, this method being known as *case hardening*. This can be done by heating the steel while it is packed in charcoal and barium carbonate, *pack carburising*, or in a furnace in an atmosphere of a carbon-rich gas, *gas carburising*, or alternatively in a bath of liquid sodium cyanide, *cyanide carburising*. These methods might, for example, result in a steel having an inner core containing 0.2% carbon and surface layers with 0.9% carbon.

Other processes that can be used involve changing the surface composition by diffusing nitrogen into it to produce hard compounds, nitrides. This is done by heating the steel in an atmosphere of ammonia gas and hydrogen. The process is known as *nitriding*. *Carbonitriding* involves heating the steel in an atmosphere containing both carbon and ammonia and allowing both carbon and nitrogen to diffuse into the surface layers.

5.6 Structure of polymers

The plastic washing-up bowl, the plastic measuring rule and the plastic cup are all examples of materials that have polymer molecules as their basis. A polymer molecule in a plastic may have thousands of atoms all joined together in a long chain. The backbones of these long molecules are chains of atoms linked together by covalent bonds. The chain backbone is usually predominantly carbon atoms and we have long molecules formed by the repetition of basic structural units formed by groups of atoms. The term *polymer* is used to indicate that a compound consists of many repeated structural units; the prefix 'poly' means many. Each repeating structural unit in the compound is called a *monomer* or *mer*. For many plastics the monomer can be deduced by deleting the prefix 'poly' from its name. Thus the plastic called polyethylene is a polymer which has ethylene as its monomer base unit. The ethylene molecule C_2H_4 has a double bond between its two carbon atoms (a result of two electrons being shared between the carbon atoms). When the polyethylene is formed, the double bond is 'opened up' and as a result the ethylene molecule has a bond free for bonding with another molecule. The 'opened up' ethylene molecules $-CH_2-CH_2-$ can thus link together to form a long chain, i.e. the polyethylene molecule. Figure 5.20 shows the form of the ethylene molecule and part of the resulting polymer chain.

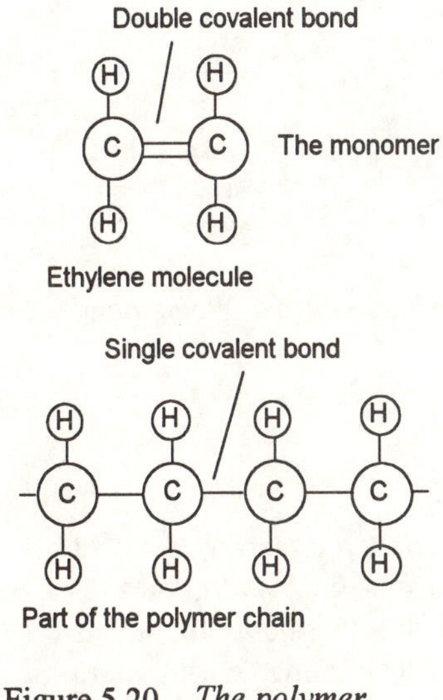

Figure 5.20 *The polymer polyethylene and its monomer*

Figure 5.20 gives a two-dimensional representation of the polymer molecule. In fact the carbon atoms do not form a straight line because the carbon atoms have a directionality which means that the angle between the bonds between a carbon atom and its two neighbouring carbon atoms is about 109°. The result is rather a zig-zag form of molecule.

5.6.1 Forms of polymer chains

The molecular chains of polyethylene are said to be linear chains. Other possible forms of polymer chains are branched and cross-linked, as illustrated in Figure 5.21 (for simplicity, only the carbon atoms are indicated).

Linear

Branched

Cross-linked

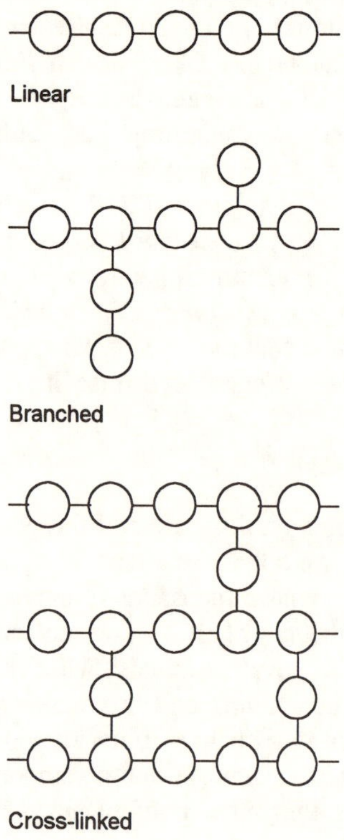

Figure 5.21 *Forms of chains*

The term *homopolymer* is used to describe those polymers that are made up of just one monomer: for instance, polyethylene is made up of only the monomer ethylene. Other types of polymers, *copolymers*, can be produced by combining two or more monomers in a single polymer chain.

A solid polymer may thus consist of linear chains arranged in some way and held together in the solid by van der Waals bonding between chains and mechanical entanglement (Figure 5.22). Van der Waals bonds are very weak bonds formed by attraction between bits of the chain which happen to have oppositely charged bits in close proximity as a result of the charges in atoms or molecules not being uniformly distributed. The linear chains have no side branches or strong bonds providing cross-links with other chains and can thus move readily past each other, breaking and remaking van der Waals bonds. If, however, the chains have side branches, there is a reduction in the ease with which chains can move past each other and so the material is more rigid. If there are ionic or covalent cross-links, a much more rigid material is produced in that the chains cannot slide past each other at all and the solid may be considered to be almost just one large cross-linked chain.

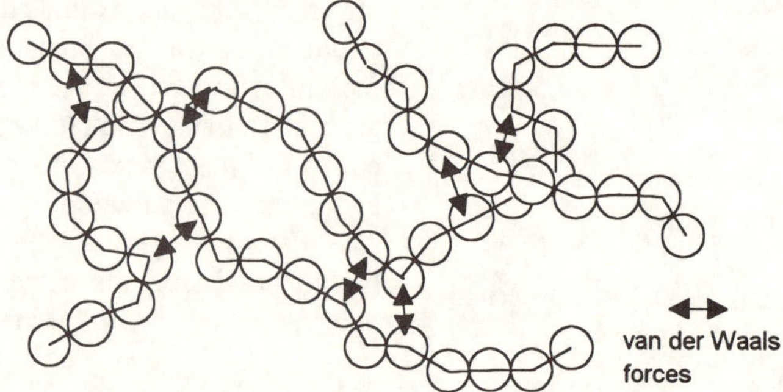

Figure 5.22 *Linear chains held together by van der Waals bonds and mechanical entanglement*

5.6.2 Classification of polymers

Polymers can be classified as *thermoplastics*, *thermosets* or *elastomers*. A simple method by which thermoplastics and thermosets can be distinguished is when heat is applied. With a thermoplastic the material softens and removal of the heat results in hardening. With a thermoset, heat causes the material to char and decompose with no softening. An elastomer is a polymer that by its structure allows considerable extensions which are reversible. Thermoplastics have linear chains or branched chains for their structure. Thermosets have a cross-linked structure. Elastomers are chains with some degree of cross-linking.

The atoms in a thermoset form a three-dimensional structure of chains with frequent cross-links between chains. The bonds linking the chains are strong and not easily broken. Thus the chains cannot slide over one another. As a consequence, thermosetting polymers are stronger and stiffer than thermoplastics. Thermoplastics offer the possibility of being heated and then pressed into the required shapes. Thermosets cannot be so manipulated. The processes by which thermosetting polymers can be shaped are limited to those where the product is formed by the chemicals being mixed together in a mould so that the cross-linked chains are produced while the material is in the mould. The result is a polymer shaped to the form dictated by the mould. No further processes, other than possibly some machining, are likely to occur.

Elastomers are polymers that can show very large, reversible strains when subject to stress. The behaviour of the material is perfectly elastic up to considerable strains, e.g. you can stretch a rubber band up to more than five times its unstrained length and it is still elastic. Elastomers have a structure consisting of tangled polymer chains which are held together by occasional cross-linked bonds. The difference between thermosets and elastomers is that with thermosets, there are frequent cross-linking bonds between chains while with elastomers there are only occasional bonds. A simple model for the elastomer structure might be a piece of very open netting. In the unstretched state the netting is in a loose pile. In the elastomer, there will be some weak temporary bonds, van der Waals bonds, between chains in close proximity to each other, these being responsible for holding the tangled chains together. When the material begins to be stretched the netting just begins to untangle itself and large strains can be produced. The van der Waals bonds between the chains will cause the elastomer to spring back to its original tangled state when the stretching forces are removed. It is not until quite large strains are applied, when the netting has become fully untangled and the structure is orderly, that the bonds between atoms in the material begin to be significantly stretched. At this point the material becomes more stiff, i.e. stress–strain graph starts to become more steep, and much larger stresses are needed to give further extensions. Figure 5.23 illustrates this.

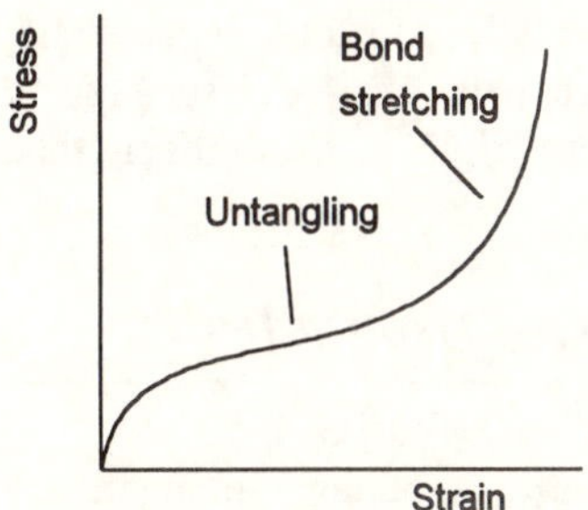

Figure 5.23 *Elastomers*

5.7 Examples of polymer structures

Thermoplastics consist of polymers with long chain molecules that are either linear chains or long chains with small branches. Linear chains have no side branches or cross-links with other chains and can be regarded as 'smooth' and 'slippery', i.e. not 'sticking' to other molecules. Because of this they can easily move past each other. If, however, the chain has branches, then there is a reduction in the ease with which chains can be made to move past each other. This shows itself in the material being more stiff, i.e. less strain produced for a given stress. Making cross-links between chains makes the chains 'sticky' and more difficult to stretch the material.

Polyethylene can be processed to have linear chains (Figure 5.20). The chains have a core of carbon atoms with hydrogen atoms attached, essentially an almost endless repetition of $-CH_2-$ units. The hydrogen atoms are small and bed into the carbon chain to give a very smooth, linear, chain. There is freedom for the chain to twist about any C–C bond and so the chain is flexible. Scaled up, the typical polyethylene

```
 H  H CH3 H  H  H  H
 |  |  |  |  |  |  |
-C -C -C -C -C -C -C-
 |  |  |  |  |  |  |
CH3 H  H  H CH3 H CH3
(a)

 H  H CH3 H  H  H CH3
 |  |  |  |  |  |  |
-C -C -C -C -C -C -C-
 |  |  |  |  |  |  |
CH3 H  H  H CH3 H  H
(b)

 H  H  H  H  H  H  H
 |  |  |  |  |  |  |
-C -C -C -C -C -C -C-
 |  |  |  |  |  |  |
 H CH3 H CH3 H CH3 H
(c)
```

Figure 5.24 *Forms of polypropylene*

chain is rather like a piece of string about 2 m long. The forces between the chains are due to the weak van der Waals bonding. The average length of the chains can be controlled and polyethylenes with different properties produced. As the length of the molecule increases so does the tensile strength, the longer molecules becoming more easily entangled and so more stress is needed to stretch the material.

We can add knobs and side branches to such a basic chain and so considerably alter the properties of the solid polymer. *Polypropylene* differs from polyethylene only to the extent that alternative carbon atoms have one of their hydrogen atoms replaced by CH_3 groups (Figure 5.24). This replacement can take a number of forms. With the *atactic* form it is random as to which side of the carbon atom the hydrogen atoms are replaced, with the *syndiatactic* form it alternates in a regular manner from one side to the other, while with the *isotactic* form all the atoms replaced are on the same side. Commercial propylene is generally predominantly isotactic. The result is knobbly chains, less slippery chains, and so a material that is more rigid and stronger than polyethylene in its linear form (Table 5.1).

Table 5.1 *Properties of linear polyethylene and polypropylene*

Polymer	Density 10^3 kg/m^3	Melting point °C	Tensile strength MPa	Tensile modulus GPa	% elongation
Linear polyethylene	0.92	115	8 to 16	0.1 to 0.3	100 to 600
Polypropylene	0.90	176	30 to 40	1.1 to 1.6	50 to 600

```
 Cl H  H  H  Cl H  Cl
 |  |  |  |  |  |  |
-C -C -C -C -C -C -C
 |  |  |  |  |  |  |
 H  H  Cl H  H  H  H
```

Figure 5.25 *PVC*

Polyvinyl chloride (PVC) has a linear chain, differing from polyethylene only to the extent that 'bulky' atoms, chlorine atoms, replace some hydrogen atoms on the chain (Figure 5.25). Commercial polyvinyl chloride is largely atactic. Because of this structure the chain is very knobbly and, when used without a plasticiser, it is a rigid and relatively hard material. Plasticisers are materials added to keep the knobbly chains apart and so permit more easily the sliding of one knobbly chain over another. Most PVC products are, however, made with a plasticiser incorporated with the polymer. The amount of plasticiser is likely to be between about 5 to 50% of the plastic, the more plasticiser added, the greater the degree of flexibility (Table 5.2).

Table 5.2 *Properties of* PVC

Amount of plasticiser	Density 10^3 kg/m^3	Tensile strength MPa	Tensile modulus GPa	% elongation
None	1.4	52 to 58	2.4 to 4.1	2 to 40
Low	1.3	28 to 42		200 to 250
High	1.2	14 to 21		350 to 450

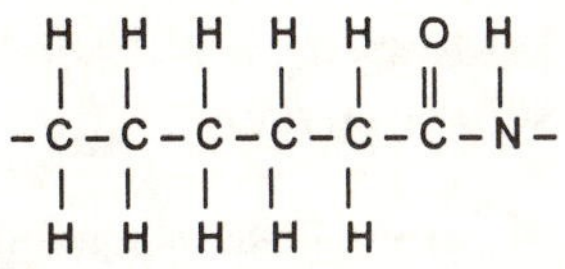

```
 H  H  H  H  H  O  H
 |  |  |  |  |  ‖  |
-C--C--C--C--C--C--N-
 |  |  |  |  |
 H  H  H  H  H
```

Figure 5.26 *The basic unit of nylon 6*

An alternative to putting knobs or branches on to a $-CH_2-$ chain is to make the chain stiffer by incorporating blocks in the backbone of the chain. An example of such a polymer is *polyethylene terephthalate (PET)*. This incorporates a six-carbon (benzene) ring in the backbone. This ring structure will not twist like the C–C bond and so the chain is stiffer. The polymer is widely used for the plastic bottles used for Coca-Cola and other drinks. It is also used as fibres for clothing, being known then as Terylene or Dacron. Polyamides, i.e. nylons, consist of amide groups of atoms separated by lengths of (CH_2) chains. The lengths of these chains can be varied to give different forms of nylon. Figure 5.26 shows the form of nylon 6, the 6 referring to the number of carbon atoms in the chain before the chain repeats itself. Nylon 6.6 has two different six-carbon length molecules stuck together. In general, nylon materials are strong, tough and have relatively high melting points (Table 5.3). They do, however, tend to absorb moisture, the effect of which is to reduce their tensile strength.

Table 5.3 *Properties of nylons*

Polymer	Density 10^3 kg/m^3	Melting point °C	Tensile strength MPa	Tensile modulus GPa	% elongation
Nylon 6	1.13	225	75	1.1 to 3.1	60 to 320
Nylon 6.6	1.1	265	80	2.8 to 3.3	60 to 300

One way of modifying the properties of a polymer is to blend two or more polymers. Polystyrene has bulky side groups attached irregularly to the polymer chain and, with no additives, is a brittle, transparent material (Table 5.4); it finds it main use as containers for cosmetics, light fittings, toys and boxes. A toughened form of polystyrene is produced by blending it with rubbers to produce high impact polystyrene (HIPS). This overcomes the problem of brittleness that occurs with polystyrene alone, giving a material which has a considerable number of uses, e.g. cups in vending machines, casings for cameras, projectors, radios, television sets and vacuum cleaners. *Acronitrile-butadiene-styrene terploymer* (ABS) is produced by forming chains with three different polymers to give a tough, stiff and abrasion resistant material which is used for such items as casings for telephones, vacuum cleaners, hair dryers, raios, television sets, luggage and food containers.

Table 5.4 *Properties of polystyrene and ABS*

	Density 10^3 kg/m^3	Tensile strength MPa	Tensile modulus GPa	% elongation
Polystyrene	1.1	35 to 60	2.5 to 4.1	1 to 3
HIPS	1.1	16 to 42	1.8 to 3.1	8 to 50
ABS	1.1	17 to 58	1.4 to 3.1	10 to 140

Example

What structural changes could be used to make a polymer material consisting of linear chains of $-CH_2-$ groups a stiffer material?

The stiffness can be increased by replacing some of the hydrogen atoms by bulky atoms or groups of atoms, introducing side branches to the chain, replacing some of the carbon atoms by groups of atoms, or introducing cross-links between polymer chains.

5.7.1 Thermosets

Many thermosets consist of small molecules which are linked together to form a highly cross-linked structure. This gives a very stiff material. *Phenolics* (Figure 5.27) are an example of such a structure.

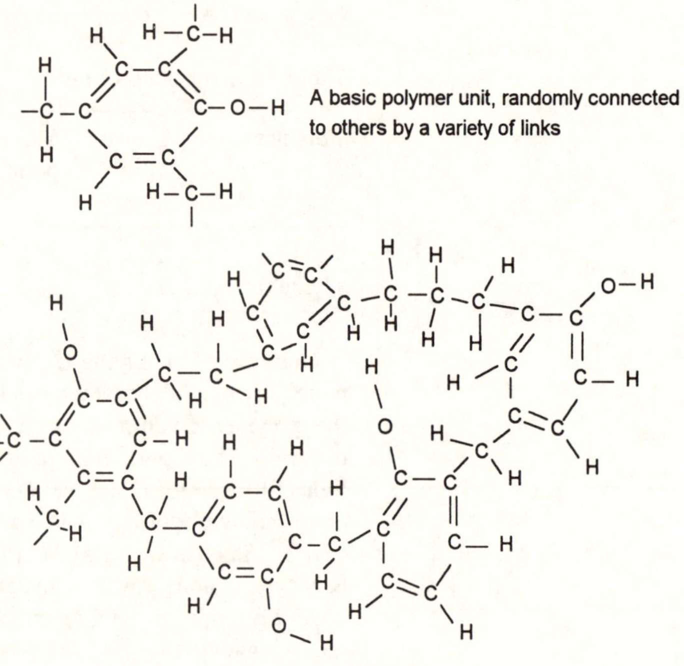

A basic polymer unit, randomly connected to others by a variety of links

Figure 5.27 *Cross-linked phenol formaldehyde*

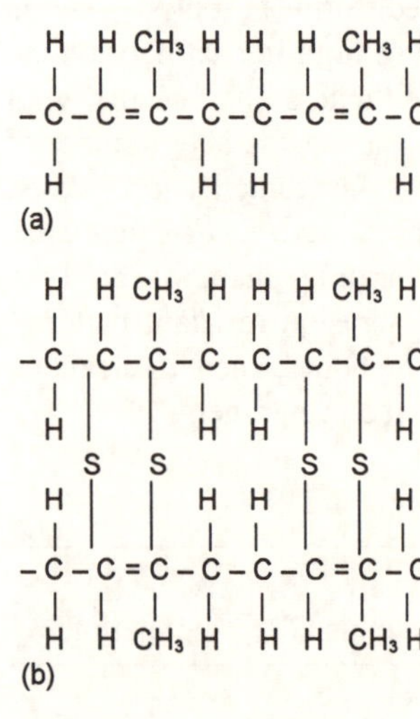

Figure 5.28 *(a) Isoprene chain, (b) isoprene chains linked by sulphur*

5.7.2 Elastomers

Many elastomers consist of long linear chains linked by small molecules. *Natural rubber* is an example of such a material. The monomer from which natural rubber polymerises is isoprene C_5H_3. The monomer links up to form long chain molecules with some 20 000 carbon atoms. When rubber is *vulcanised*, molecules of sulphur form cross-links between chains (Figure 5.28). The sulphur breaks some of the double bonds between carbon atoms in the chains to give sulphur cross-links. The amount of sulphur added determines the amount of cross-links and hence the properties of the rubber. The greater the number of cross-links, the harder it is to stretch the rubber. The rubber of a rubber band has typically about one sulphur cross-link every few hundred carbon atoms.

The isoprene long chain molecule can exist in two different forms. In one form, referred to as the *cis structure*, the CH_3 groups are all on the same side of the chain (as in Figure 5.28). This concentration of the CH_3 groups all on one side of the chain allows the chain to bend easily and coil in a direction which puts the CH_3 groups on the outside of the bend. In the other form, the *trans structure*, the CH_3 groups alternate between opposite sides of the chain. This has the result that the chain cannot easily bend since the CH_3 groups get in the way. Cis-polyisoprene is natural rubber and shows a high degree of flexibility. Trans-polyisoprene is gutta-percha and is inflexible in comparison with the cis form.

5.8 Polymer crystallinity

A *crystalline* structure is one in which there is an orderly arrangement of particles; a structure in which the arrangement is completely random is said to be *amorphous*. Many polymers are amorphous with the polymer chains being completely randomly arranged in the material. Figure 5.29 illustrates this, the chains being shown as lines, individual atoms not being indicated. Linear polymer molecules can, however, assume an arrangement which is, at least partially, orderly. The long molecules fold backwards and forwards on themselves to give a concertina-like system, each loop being approximately 100 carbon atoms long (Figure 5.30). The arrangement is said to be *crystalline*.

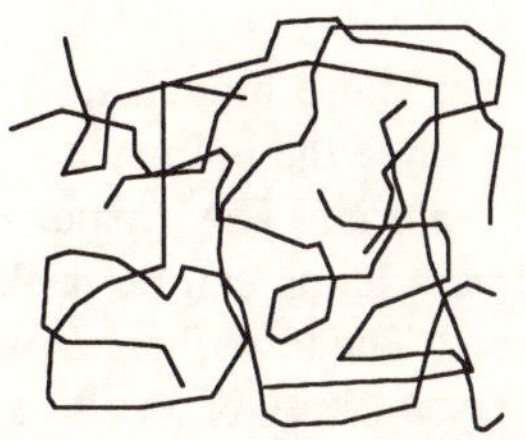

Figure 5.29 *Amorphous polymer*

Often when a polymer is cooling from the melt, a number of such platelets grow with regions of amorphous material between them. As with the growth of metal and other crystals, growth from the melt occurs outwards from initiating nuclei. The platelets grow outwards, trapping amorphous material between them. This amorphous material wedges the platelets apart and they continue growing outwards until they are bent back on themselves and touch, thus forming a sheaf-like sphere called a *spherulite* (Figure 5.31).

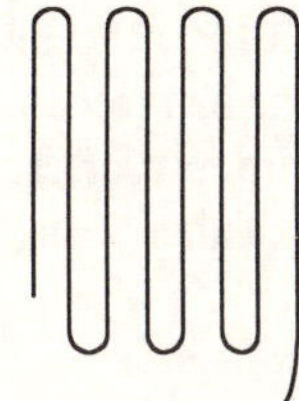

Figure 5.30 *Folded chain*

The tendency of a polymer to crystallise is determined by the form of the polymer chains, Figure 5.32 graphically illustrating the problem. Linear polymers can crystallise to quite an extent, complete crystallisation is not, however, obtained in that there is invariably some regions of disorder. For example, linear polyethylene chains can have some 95% of the material crystalline. PVC is essentially just the polyethylene molecule with some of its hydrogen atoms replaced by chlorine atoms to give a knobbly structure for the molecule. The molecule does not, however, give rise to a crystalline structure because the chlorine atoms are rather bulky and not regularly spaced along the chain and so get in the way of orderly packing. Polypropylene has a molecule rather like that of polyethylene but with some of the hydrogen atoms replaced by CH_3 groups. These are, however, regularly spaced along the molecular chain and thus some degree of orderly packing and hence crystallinity is possible.

Figure 5.31 *A spherulite*

Table 5.5 shows the forms of molecular chains and degree of crystallinity possible for some common polymers. Polymers with side branches show less tendency to crystallise since the branches get in the

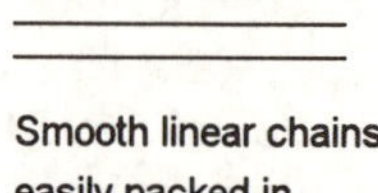

Smooth linear chains easily packed in an orderly manner

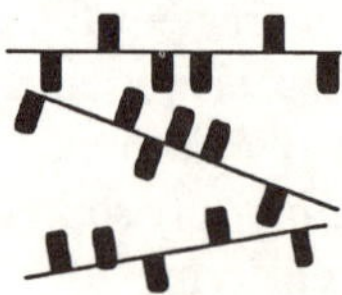

Irregular knobbly chains are not orderly packed

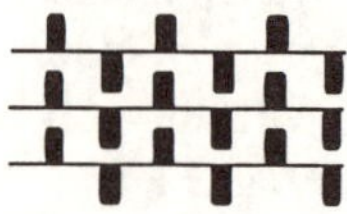

Regular knobbly chains can be orderly packed but it is more difficult

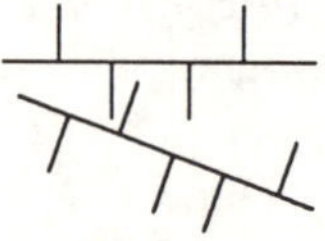

Irregularly branched chains are not orderly packed

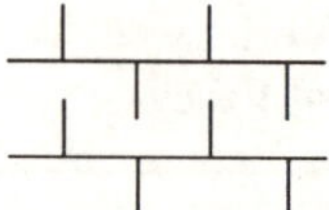

Regularly branched chains can be orderly packed but it is more difficult

Figure 5.32 *Packing chains*

way of the orderly arrangement. If the branches are completely regularly spaced along the chain then some crystallinity is possible; irregularly spaced branches make crystallinity improbable.

Table 5.5 *Crystallinity of some common polymers*

Polymer	Form of chain	% possible crystallinity
Polyethylene	Linear	95
	Branched	60
Polypropylene	Regularly spaced groups on linear chain	60
Polyvinyl chloride	Irregularly spaced bulky chlorine atoms on linear chain	0
Polystyrene	Irregularly spaced bulky side groups on linear chain	0

Polyethylene can be produced in a linear form and a branched form, the linear form showing about 95% crystallinity while the branched form might show crystallinity in about 50% of the material. The greater the crystallinity of a polymer, the closer the polymer chains can be packed and so the greater the density of the solid polymer. The term *high density polyethylene* is used for the polyethylene with crystallinity of about 95%, *low density polyethylene* for that with crystallinity of about 50%. The high density polyethylene has a density of about 950 kg/m^3, the low density about 920 kg/m^3. The closer packing of the chains in the high density polyethylene means that there can be more inter-chain van der Waals bonds, hence a higher melting point (138°C) than the low density form (115°C). The more such bonds the more stress has to be applied to produce a particular strain and so the greater the amount of crystallinity the higher the tensile modulus and tensile strength (Table 5.6).

Table 5.6 *Effect of crystallinity on polyethylene properties*

Polymer	% crystallinity	Modulus GPa	Strength MPa
Polyethylene	95	21 to 38	0.4 to 1.3
Polyethylene	60	7 to 16	0.1 to 0.3

When an amorphous polymer is heated, it shows no definite melting temperature but progressively becomes less rigid. This is because the arrangement of the chains in the solid is disorderly, just like in a liquid, and so there is no structural change occurring at melting. With a crystalline polymer, there is an abrupt change in structure at a particular temperature when the crystalline structure changes to a disorderly structure. If the density of the polymer were being monitored, there would be an abrupt change in density when this occurs as a result of a change in the way the chains are packed together. This temperature is termed the *melting point*.

5.8.1 Glass transition temperature

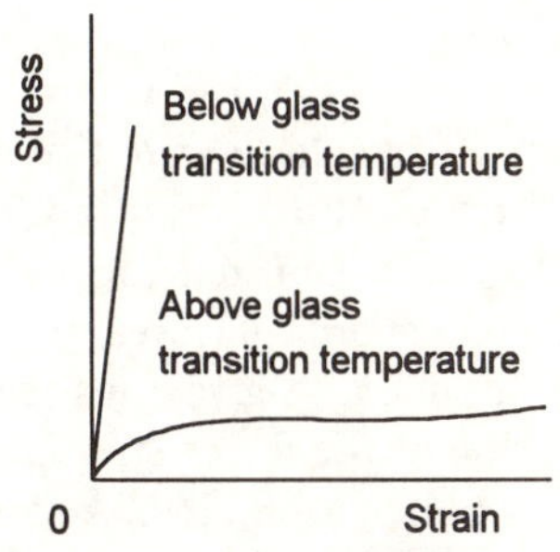

Figure 5.33 *Stress–strain graphs below and above glass transition temperature*

PVC, without any additives, at room temperature is a rather rigid material. It is often used in place of glass. But if it is heated to a temperature of about 87°C a change occurs, the PVC becomes flexible and rubbery. The PVC below this temperature gives only a moderate elongation before breaking, above this temperature it stretches a considerable amount. If amorphous polymers are heated, there is a temperature at which they change from being a stiff, brittle, glass-like material to a rubbery material. This temperature is called the *glass transition temperature* T_g. Below this temperature, segments within the molecular chains are unable to move and the material is stiff with a high elastic modulus and generally rather brittle. Above this temperature, there is sufficient thermal energy for some motion of segments of the chains to occur. The material then becomes less stiff with a lower elastic modulus and more like an elastomer and rubbery (Figure 5.33).

Perspex is an example of a glassy polymer at room temperature. It has a glass transition temperature of 120°C. Thus when it is heated above this temperature it can be easily bent and twisted. Below that temperature it is much stiffer and fairly brittle. This property can be used to enable such polymers to be moulded into shapes required for products.

Below the glass transition temperature only a very limited molecular motion is possible; above the glass transition temperature quite a large amount of motion is possible. The extent to which motion is possible at any particular temperature depends on the structure of the polymer molecules and how well they can move past each other. Thus linear chain molecules tend to have lower glass transition temperatures than molecules with bulky side groups or branches and these, in turn, have lower values than cross-linked polymers. The greater the degree of linking the higher the glass transition temperature. Table 5.7 shows some typical values.

Table 5.7 *Glass transition temperatures*

Material		T_g°C
Thermoplastic:	Polyethylene, low density	–90
	Polypropylene	–27
	Polyvinyl chloride	+80
	Polystyrene	+100
Thermoset:	Phenol formaldehyde	Decomposes first
	Urea formaldehyde	Decomposes first
Elastomer:	Natural rubber	–73
	Butadiene styrene rubber	–58
	Polyurethane	–48

In compounding a plastic, other materials are added to the polymer. These can affect the glass transition temperature, e.g. a plasticiser depresses the glass transition temperature by coming between the polymer molecules and weakening the forces between them.

Activity
List the various methods that can be used to change the properties of polymers, giving the changes that occur and the reasons.

5.9 Stretching polymers

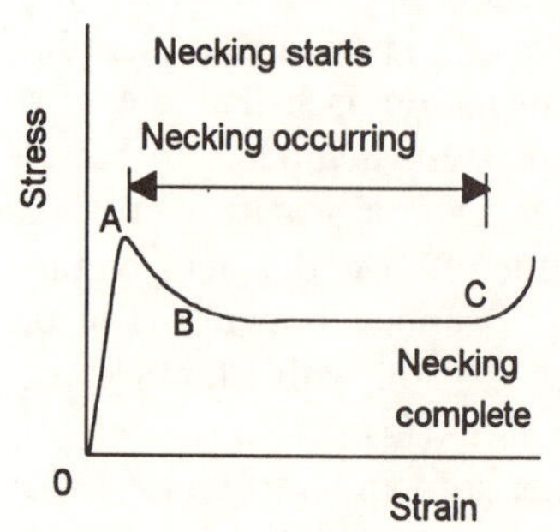

Figure 5.34 *Stress–strain graph for a crystalline polymer*

Consider what happens with a crystalline thermoplastic when it is stretched. Figure 5.34 shows the typical form of stress–strain graph. When stress is applied, the first thing that begins to happen is that there is some movement of folded chains past each other. However, when point A is reached the polymer chains start to unfold to give a material with the chains all lined up along the direction of the forces stretching the material. The material shows this by starting to exhibit *necking* (Figure 5.35), i.e. a section of the material suddenly shows a marked contraction in its cross-section. As the stress is further increased, the necking spreads along the material with more and more chains unfolding. Eventually, when the entire material is at the necked stage, all the chains have lined up. The material is said to be *cold drawn*. Such a material has, as a result of the orientation of the molecular chains, different properties to the undrawn material. The material is stiffer, i.e. tensile modulus is higher, and stronger. Typically, with polyethylene, the tensile modulus increases from about 1 GPa to 10 GPa and the tensile strength from about 30 MPa to perhaps 200 MPa. However, the percentage elongation is reduced, typically from about a few hundred per cent to less than 10%.

The above sequence of events only tends to occur if the material is stretched slowly and sufficient time elapses for the molecular chains to unfold. If a high strain rate is used, the material is likely to break without the chains all becoming lined up. The plastic used for making polythene bags is a crystalline polymer. Try cutting a strip of polythene from such a bag and pulling it between your hands and see the necking develop with low rates of strain. Try quickly breaking a strip of polyethylene before orientating the molecules and then another strip after it has been stretched and the molecules orientated, the difference in tensile strength should be apparent.

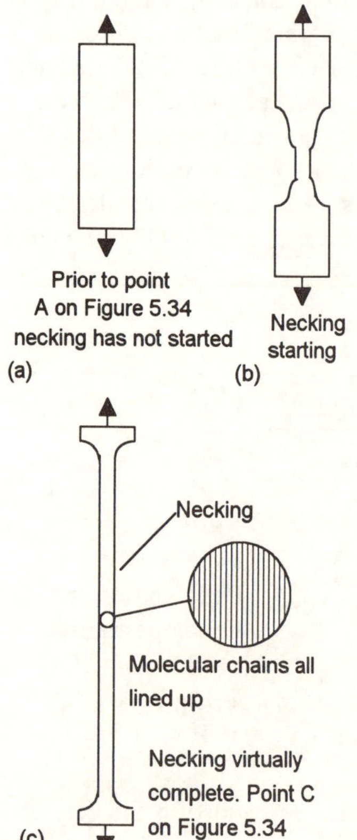

Figure 5.35 *Necking*

In order to improve the strength of polymer fibres, e.g. polyester fibres, they are put through a drawing operation to orientate the polymer chains. Stretching a polymer film causes orientation of the polymer chains in the direction of the stretching forces. The result is an increase in strength and stiffness in the stretching direction. Such stretching is referred to as *uniaxial orientation*. The material is, however, weak if forces are applied in directions other than the stretching forces and has a tendency to split. For the polyester fibres this does not matter as the forces will be applied along the length of the fibre; with film this could be a series defect. The problem can be overcome by using a biaxial stretching process in which the film is stretched in two directions at right angles to each other. The film has then *biaxial orientation*.

If orientated polymers are heated to above their glass transition temperatures, they lose their orientation. On cooling they are no longer orientated. This is made use of with *shrinkable films*. The polymer film is stretched and becomes longer and orientated. If it is then wrapped around some package and heated, the film loses its orientation and contracts back to its prestretched state. The result is a plastic film tightly fitting the package.

Crystalline polymers are used up to their melting temperature. They can be hot formed and shaped at temperatures above the melting point or cold formed and shaped at temperatures between the glass transition temperature and the melting point.

Activity

Cut a strip of polythene about 200 mm by 10 mm from a polythene bag; the edges should be clean and not ragged. Now pull the strip slowly and observe the results and your feelings as to how much force is needed for the stretching. Continue the stretching until the material breaks. Write notes describing the behaviour of the polyethylene.

5.9.1 Stretching amorphous polymers

Below the glass transition temperature, an amorphous polymer is glass-like and rather stiff and brittle. This is because, when so cold, no chains or parts of chains can move. If the temperature is increased to above the glass transition temperature, the material behaves in a rubbery fashion. This polymer is then very flexible, i.e. a much lower value of elastic modulus, and is able to withstand large and recoverable strains, i.e. just like a rubber band. This is because there is now sufficient thermal energy being supplied for not only side groups on chains to be able to rotate but also entire segments of the chain to rotate and move.

Figure 5.36 shows how the modulus of elasticity varies for such a material. Note that the modulus is plotted on a scale where each scale marking sees the value increase by a power of ten.

Amorphous polymers tend to be used below their glass transition temperature. They are, however, formed and shaped at temperatures above the glass transition temperature; they are then in a soft condition.

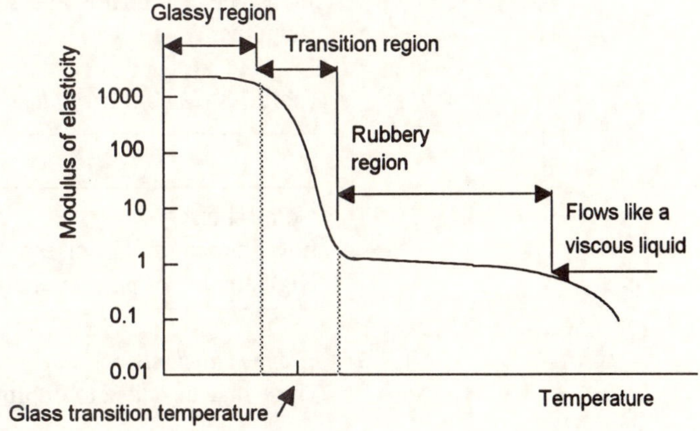

Figure 5.36 *Effect of temperature on the modulus of elasticity of an amorphous polymer*

Elastomers are, under normal conditions, amorphous polymers that at room temperature are above their glass transition temperatures and so exhibit rubbery behaviour. However, if you cool a rubber sufficiently it becomes brittle and shows glassy behaviour. Ordinary rubber tubing if cooled in liquid nitrogen shows such behaviour. Most polymers become rubbery at some temperature, the exception being heavily cross-linked thermosets, which decompose before they reach their glass transition temperatures.

5.10 Additives with polymers

The term *plastic* is commonly used to describe materials based on polymers. Such materials, however, invariably contain other substances that are added to the polymers to give the required properties. The following are some of the main types of additives:

1 *Stabilisers*

Since some polymers are damaged by ultraviolet radiation, protracted exposure to the sun can lead to a deterioration of mechanical properties. An ultraviolet absorber is thus often added to the polymer, such an additive being called a *stabiliser*. Carbon black is often used for this purpose.

2 *Plasticisers*

These are added to a polymer to make it more flexible. In one form this may be a liquid which is dispersed throughout the solid, filling the space between the polymer chains and acting like a lubricant and permitting the chains to more easily slide past each other. This is termed *external plasticisation*. The plasticiser decreases the crystallinity of polymers as it tends to hinder the formation of orderly arrays of polymer chains; it also reduces the glass transition temperature. Table 5.8 shows the effect of plasticiser on the properties of PVC. *Internal plasticisation* involves modifying the polymer chain by the introduction into it of bulky side groups. These force the polymer chains further apart, thus reducing the attractive forces between chains and so permitting easier flow of chains past each other.

Table 5.8 *The effect of plasticiser on PVC properties*

	Tensile strength MPa	% elongation
No plasticiser	52 to 58	2 to 40
Low amount of plasticiser	28 to 42	200 to 250
High amount of plasticiser	14 to 21	350 to 450

3 *Flame retardants*

These may be added to improve fire-resistant properties.

4 *Pigments and dyes*

These are added to give colour to the material.

5 *Fillers*

The properties and cost of a plastic can be markedly affected by the addition of substances termed *fillers*. Since fillers are generally cheaper than the polymer, the overall cost of the plastic is reduced. Up to 80% of a plastic may be filler. Examples of fillers are glass fibres to increase the tensile strength and impact strength, mica to improve electrical resistance, graphite to reduce friction, wood flour to increase tensile strength. One form of additive used is a gas to give foamed or expanded plastics. Expanded polystyrene is used as a lightweight packaging material, foamed polyurethane as a filling for upholstery.

6 *Lubricants and heat stabilisers*

These may be added to assist the processing of the material.

5.11 Ceramics

The bonds between the atoms in ceramics are ionic or covalent. Because such bonds are strong, ceramics have high melting points. Examples of ionic bonded ceramics are magnesium oxide MgO, alumina Al_2O_3 and zirconia ZrO_2. Ionic bonded materials are held together by the strong forces between oppositely charged ions. Figure 5.37 shows the type of structure occurring with MgO, it having positively charged magnesium ions and negatively charged oxygen ions. Magnesium oxide is an engineering ceramic which is used as a refractory in furnaces. Ceramics which are predominantly covalent bonded are combinations of two non-metals like silica and oxygen or, sometimes, just pure elements like diamond with just carbon atoms. Covalent bonded materials involve neighbouring atoms sharing electrons. Diamond (Figure 5.38(a)) is an example of a simple covalent bonded ceramic, all the bonded atoms being carbon. The engineering ceramic silicon carbide SiC (Figure 5.38(b)) has a structure similar to that of diamond with some of the carbon atoms being replaced by silicon.

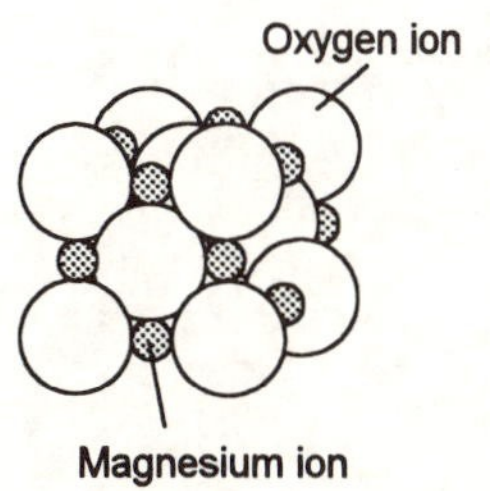

Figure 5.37 *Magnesium oxide*

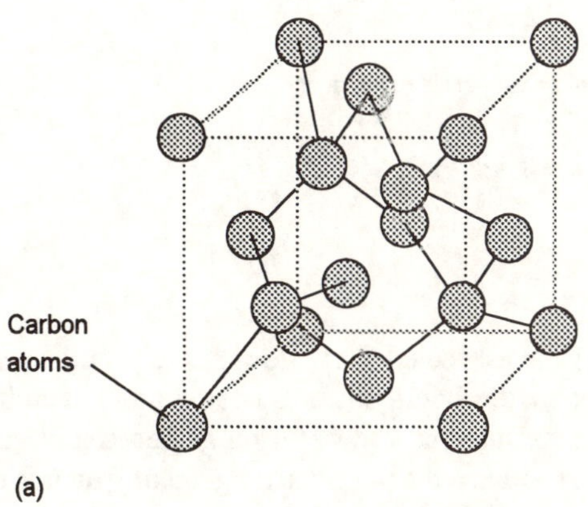

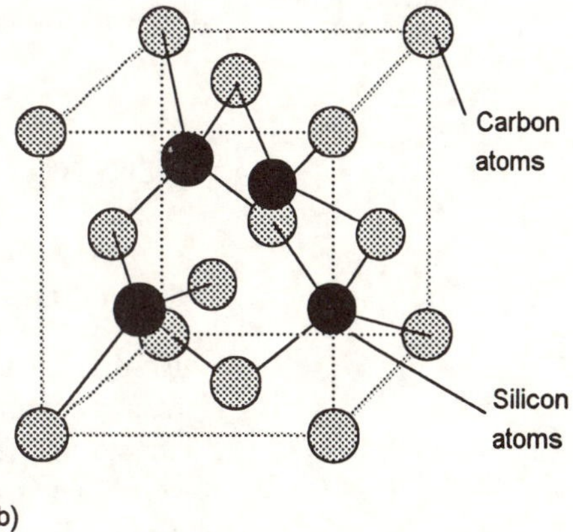

Figure 5.38 *(a) Diamond, (b) silicon carbide*

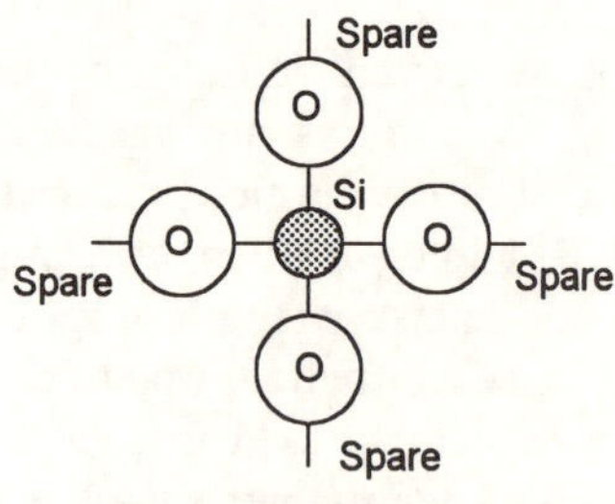

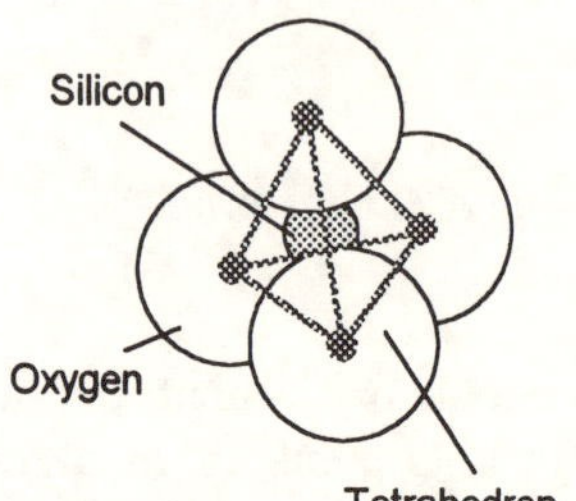

Figure 5.39 *The silicon–oxygen tetrahedron*

Silica forms the basis of a large variety of ceramics. A silicon atom forms covalent bonds with four oxygen atoms to give a tetrahedron-shaped structure (Figure 5.39). This form of structure leaves the oxygen atoms with 'spare' bonds with which to link up with other silicon–oxygen tetrahedra or metal ions. Crystalline silica can be described as a number of the silicon–oxygen tetrahedra joined together in an orderly manner corner to corner to give three-dimensional structure. Quartz is such a crystalline silica structure (Figure 5.40).

Other silica structures can be produced by linking tetrahedra in chains and then linking adjacent chains by metallic ions (Figure 5.41). The links in the chain are covalent bonds while the links between adjacent chains are ionic. The tetrahedra can also link up to give sheets instead of chains. Such a sheet is the basis of many minerals, e.g. clay. When water is added to clay, the water molecules attach by van der Waals bonds and so the sheets can glide easily over each other. As a consequence the clay is plastic and readily mouldable.

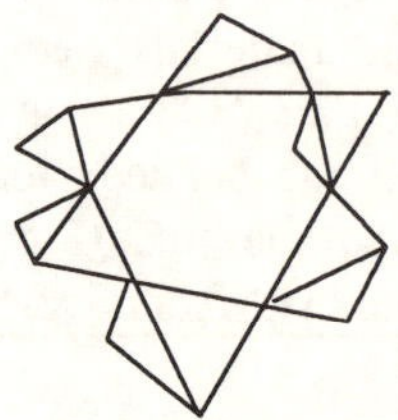

Figure 5.40 *Crystalline quartz*

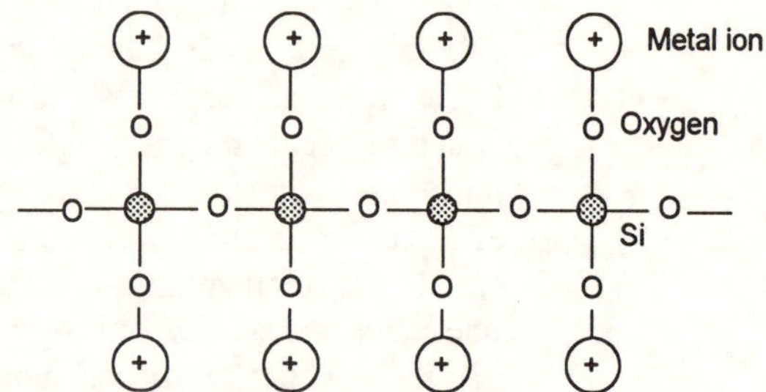

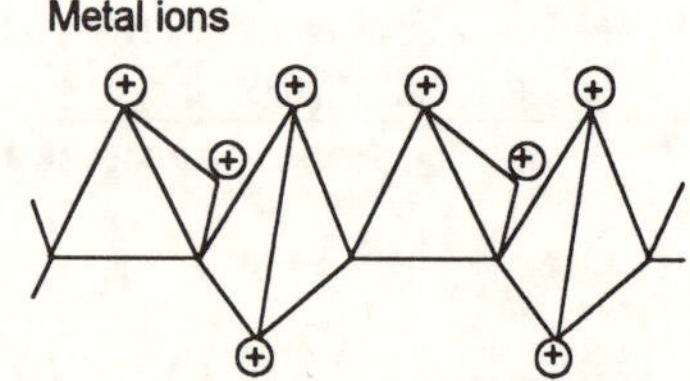

Figure 5.41 *Single chain structure*

5.11.1 Glasses

If silica in the liquid state is cooled very slowly it crystallises at the freezing point. However, if the liquid silica is cooled more rapidly it is unable to get all its atoms into the orderly arrangement required of a crystal and the resulting solid is a disorderly arrangement which is called a *glass*. Figure 5.42 shows such a structure.

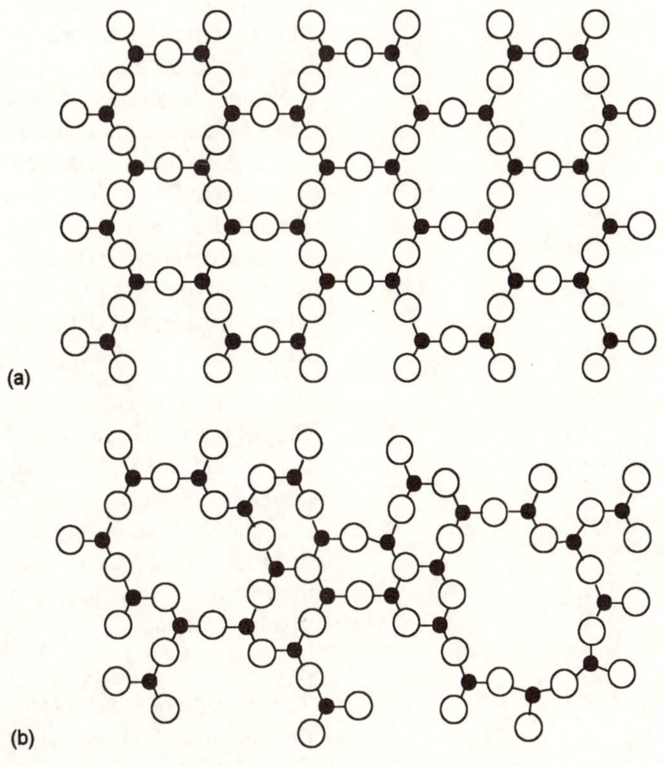

Figure 5.42 *Two-dimensional representations: (a) the orderly arrangement for a crystal, (b) the disorderly arrangement for a glass*

5.12 Composites

Composites are composed of two different materials bonded together. There are many examples of composite materials encountered in everyday products. Composites can be classified into four categories:

1 *Fibre reinforced*
The fibres may be continuous throughout the matrix or short fibres, and aligned in all the same direction or randomly arranged (Figure 5.43). The main functions of the fibres in a composite are to carry most of the load applied to the composite and provide stiffness. For this reason, fibre materials have high tensile strength and a high elastic modulus. Ceramics are frequently used for the fibres in composites. Ceramics have high values of tensile strength and tensile modulus, the useful asset of low density, but are brittle and the presence of quite small surface flaws can markedly reduce the tensile strength. By incorporating such fibres in a ductile matrix it is possible to form a composite which makes use of the high strength–high elastic modulus properties of the fibres and the protective properties of the matrix material to give a composite with properties considerably better than with just the matrix material alone or the properties of damaged fibre material. For example, there are

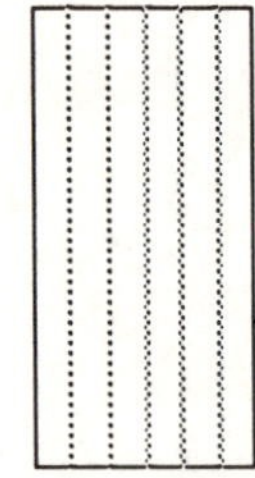
(a) Continuous, aligned

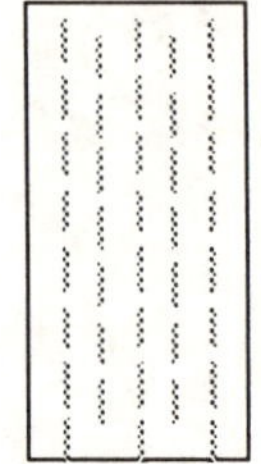
(b) Discontinuous, aligned

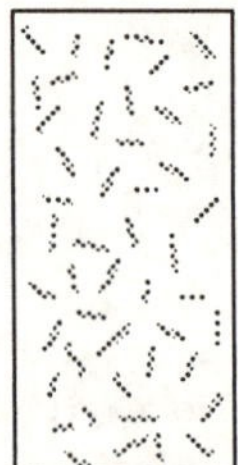
(c) Discontinuous, random

Figure 5.43 *Forms of fibre reinforcements*

composites involving glass or carbon fibres in polymers, ceramic fibres in metals and metal fibres in ceramics. Many plastics are glass fibre reinforced, the result being a much stronger and stiffer material than given by the plastic alone. Polymer composites are used for such applications as car instrument panels, domestic shower units and crash helmets. Vehicle tyres are rubber reinforced with woven cords. A common example of a metal-reinforced ceramic composite is reinforced concrete. The composite material enables loads to be carried that otherwise could not have been carried by the concrete alone. Ceramic-reinforced metal composites are used for rocket nozzles, wire-drawing dies, cutting tools and other applications where hardness and performance at high temperatures might be required.

2 *Particle reinforced*
Cermets are composites involving ceramic particles in a metal matrix and are widely used for the tips of cutting tools. Ceramic particles are hard but brittle and lack toughness; the metal is soft and ductile. Embedding the ceramic particles in the metal gives a material that is strong, hard and tough. Glass spheres are widely used with polymers to give a composite which is stronger and stiffer than the polymer alone.

3 *Dispersion strengthened*
The strength of a metal can be increased by small particles dispersed throughout it. One way of introducing a dispersion of small particles throughout a metal uses sintering. This process involves compacting a powdered metal powder in a die and then heating it to a temperature high enough to knit together the particles in the powder. If this is done with aluminium the result is a fine dispersion of aluminium oxide, about 10%, throughout an aluminium matrix. The aluminium oxide occurs because aluminium in the presence of oxygen is coated with aluminium oxide and when the aluminium powder is compacted, much of the surface oxide film becomes separated from the aluminium and becomes dispersed through the metal. The aluminium oxide powder, a ceramic, dispersed throughout the aluminium matrix gives a stronger material than that which would have been given by the aluminium alone. At room temperature, the tensile strength of the sintered aluminium powder (SAP) is about 400 MPa, compared with that of about 90 MPa for the aluminium.

4 *Laminates*
Laminates are composites in which materials are sandwiched together. *Plywood* is an example of a laminated material. It is made by gluing together thin sheets of wood with their grain directions at right angles to each other (Figure 5.44). The grain directions are the directions of the cellulose fibres in wood, a natural composite, and thus the resulting structure, the plywood, has fibres in mutually perpendicular directions. Thus, whereas the thin sheet had properties that were directional, the resulting laminate has no such

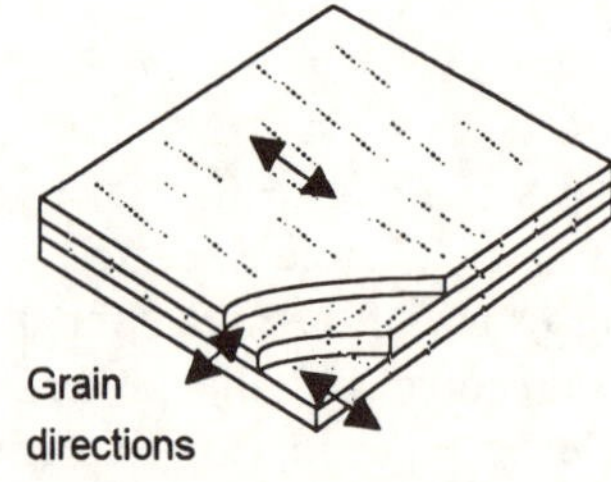

Figure 5.44 *Plywood, 3-ply*

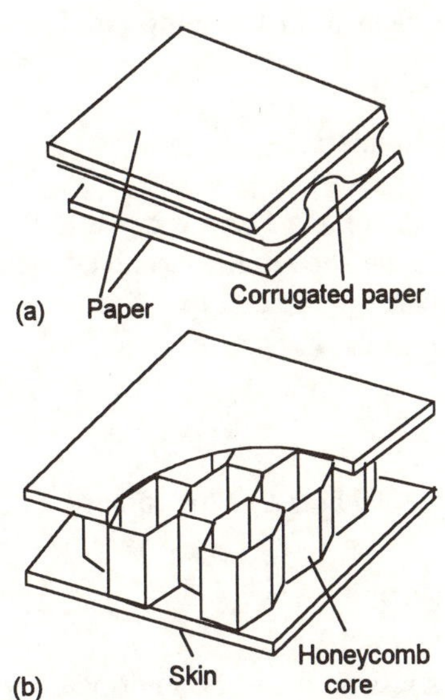

Figure 5.45 *(a) Corrugated cardboard, (b) honeycomb structure*

directionality. *Corrugated cardboard* is another form of laminated structure (Figure 5.45(a)), consisting of paper corrugations sandwiched between layers of paper. The resulting structure is much stiffer, in the direction parallel to the corrugations, than the paper alone. A structure with a different sandwiched core has an aluminium, polymer or paper *honeycomb structure* (Figure 5.45(b)) sandwiched between thin sheets of metal or polymer. Such a structure has good stiffness and is very light and is often used for structural panels.

Activity

Investigate the properties of a sheet of cardboard and a sheet of corrugated carboard, considering bending in various directions, and write a note explaining how the corrugated cardboard form of laminate improves the properties of just plain cardboard.

5.12.1 Fibres in a matrix

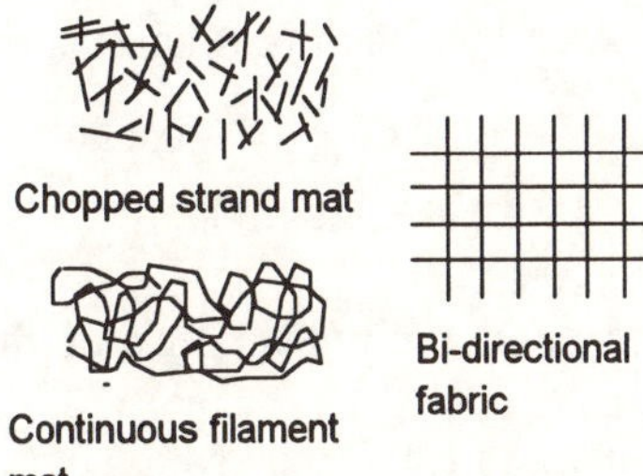

Figure 5.46 *Mats and fabrics*

Glass fibres are widely used for reinforcing polymers. The fibres may be long lengths running through the length of the composite, or discontinuous short lengths randomly orientated within the composite. Another form of composite uses glass fibre mats or cloth in the plastic. These may be in the form of randomly orientated fibre bundles which may be chopped or continuous and are loosely held together with a binder or a woven fabric (Figure 5.46). The effect of the fibres is to increase both the tensile strength and tensile modulus, the amount of change depending on both the form the fibres take and the amount. The continuous fibres give the highest tensile strength and tensile modulus composite but with a high directionality of properties. The strength along the direction of the fibres could be as high as 800 MPa while that at right angles to the fibre direction may be as low as 30 MPa, i.e. just about the strength of the plastic alone. Randomly orientated short fibres do not lead to this directionality of properties and do not give such high strengths and tensile modulus. Table 5.8 gives examples of the strength and modulus values obtained with glass fibre-reinforced polymer.

Table 5.8 *Properties of reinforced polyester*

Composite	% by weight of glass fibre	Tensile modulus GPa	Tensile strength MPa
Polyester alone		2 to 4	20 to 70
With short random fibres	10 to 45	5 to 14	40 to 180
With plain weave cloth	45 to 65	10 to 20	250 to 350
With long fibres	50 to 80	20 to 50	400 to 1200

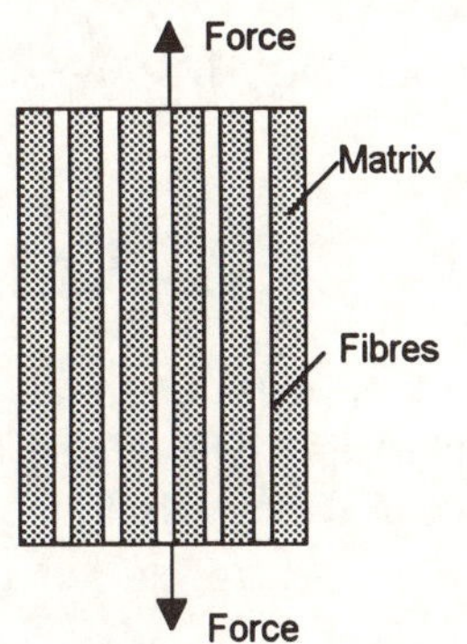

Figure 5.47 *Continuous fibres in a matrix*

5.12.2 Continuous fibres in a matrix

Consider a composite made up of continuous fibres, all parallel to the loading direction, in a matrix (Figure 5.47). When tensile forces are applied to the composite rod, then each element in the composite has a share of the applied forces. Thus:

total force = forces on fibres + force on matrix

But since stress = force/area, then the force on the fibres is equal to the product of the stress σ_f on the fibres and their total cross-sectional area A_f. Likewise, the force on the matrix is equal to the product of the stress σ_m on the matrix and its cross-sectional area A_m. Hence:

$$\text{total force} = \sigma_f A_f + \sigma_m A_m$$

Dividing both sides of the equation by the total area A of the composite:

$$\text{stress on composite} = \frac{\text{total force}}{\text{total area}} = \sigma_f \frac{A_f}{A} + \sigma_m \frac{A_m}{A}$$

Thus the stress on the composite is the stress on the fibres multiplied by the fraction of the area that is fibres plus the stress on the matrix multiplied by the fraction of the area that is matrix.

Suppose we have glass fibres with a tensile strength of 1500 MPa in a matrix of polyester with a tensile strength of 45 MPa. If the fibres occupy, say, 60% of the cross-sectional area of the composite, then the above equation indicates that the tensile strength of the composite, i.e. the stress the composite can withstand when both the fibres and matrix are stressed to their limits, will be:

$$\text{strength of composite} = 1500 \times 0.6 + 45 \times 0.4 = 918 \text{ MPa}$$

The composite has a much higher tensile strength than that of the polyester alone.

If the fibres are firmly bonded to the matrix, then the elongation of the fibres and matrix must be the same and equal to that of the composite as a whole. Thus:

strain on composite = strain on fibres = strain on matrix

Dividing the stress equation above by this strain gives, since stress/strain is the tensile modulus,

$$\text{modulus of composite} = E_f \frac{A_f}{A} + E_m \frac{A_m}{A}$$

Suppose we have glass fibres with a tensile modulus of 76 GPa in a matrix of polyester having a tensile modulus of 3 GPa. If the fibres occupy, say, 60% of the cross-sectional area of the composite then the tensile modulus of the composite is:

$$\text{modulus of composite} = 76 \times 0.6 + 3 \times 0.4 = 46.8 \text{ GPa}$$

The composite has a tensile modulus much higher than that of the polyester alone.

Example
A column of reinforced concrete has steel reinforcing rods running through the entire length of the column and parallel to the axis of the column. If the concrete has a modulus of elasticity of 20 GPa and the steel 210 GPa, what is the modulus of elasticity of the column if the steel rods occupy 10% of the cross-sectional area?

$$\text{Modulus} = E_f \frac{A_f}{A} + E_m \frac{A_m}{A} = 210 \times 0.1 + 20 \times 0.9 = 39 \text{ GPa}$$

Example
Carbon fibres with a tensile modulus of 400 GPa are used to reinforce aluminium with a tensile modulus of 70 GPa. If the fibres are long and parallel to the axis along which the load is applied, what is the tensile modulus of the composite when the fibres occupy 50% of the composite area?

$$\text{Modulus} = E_f \frac{A_f}{A} + E_m \frac{A_m}{A} = 400 \times 0.5 + 70 \times 0.5 = 235 \text{ GPa}$$

5.13 Electrical conductivity

In terms of their electrical conductivity, materials can be grouped into three categories, namely conductors, semiconductors and insulators. Conductors have electrical conductivities of the order of 10^6 S/m, semiconductors about 1 S/m and insulators 10^{-10} S/m. Conductors are metals with insulators being polymers or ceramics. Semiconductors include silicon, germanium and compounds such as gallium arsenide.

In discussing electrical conduction in materials, a useful picture is of an atom as consisting of a nucleus surrounded by its electrons. The electrons are bound to the nucleus by electric forces of attraction. The force of attraction is weaker the further an electron is from the nucleus. The electrons in the furthest orbit from the nucleus are called the valence electrons since they are the ones involved in the bonding of atoms together to form compounds.

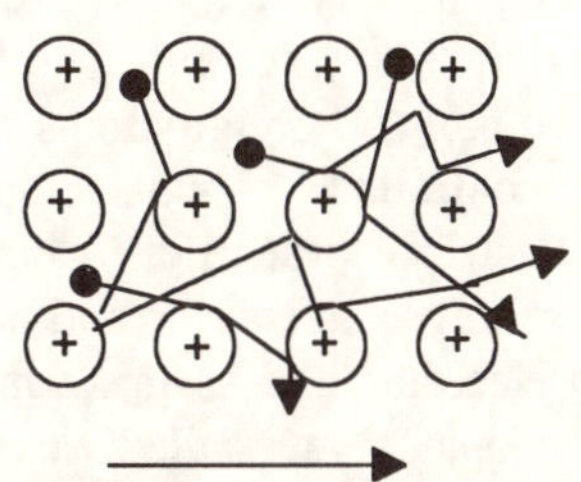

Figure 5.48 *Electric current with a metal*

Metals can be considered to have a structure of atoms with valence electrons which are so loosely attached that they drift off and can move freely between the atoms. Typically a metal will have about 10^{28} free electrons per cubic metre. Thus, when a potential difference is applied across a metal, there are large numbers of free electrons able to respond and give rise to a current. We can think of the electrons pursuing a zigzag path through the metal as they bounce back and forth between atoms (Figure 5.48). An increase in the temperature of a metal results in a decrease in the conductivity. This is because the temperature rise does not result in the release of any more electrons but causes the atoms to vibrate and scatter electrons more, so hindering their progress through the metal.

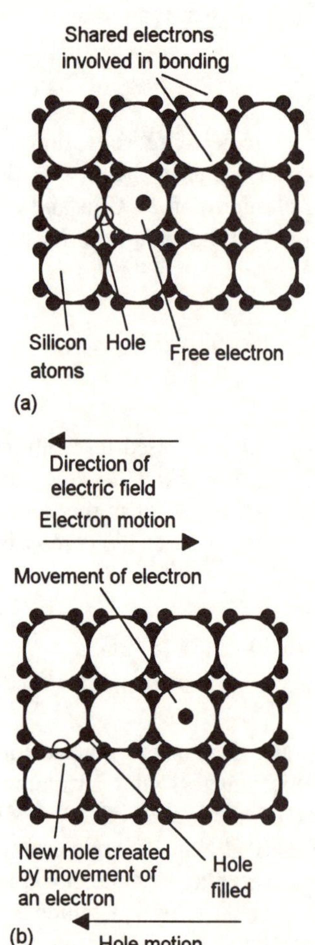

Figure 5.49 *(a) Holes and electrons in silicon, (b) movement of electrons and holes when an electric field is applied*

Insulators, however, have a structure in which all the electrons are tightly bound to atoms. Thus there is no current when a potential difference is applied because there are no free electrons able to move through the material. To give a current, sufficient energy needs to be supplied to break the strong bonds which exist between electrons and insulator atoms. The bonds are too strong to be easily broken and hence normally there is no current. A very large temperature increase would be necessary to shake such electrons from the atoms.

Semiconductors can be regarded as insulators at a temperature of absolute zero. However, the energy needed to remove an electron from an atom is not very high and at room temperature there has been sufficient energy supplied for some electrons to have broken free. Thus the application of a potential difference will result in a current. Increasing the temperature results in more electrons being shaken free and hence an increase in conductivity. At about room temperature, a typical semiconductor will have about 10^{16} free electrons per cubic metre and 10^{16} atoms per cubic metre with missing electrons.

Silicon is a covalently bonded solid with, at absolute zero, all the outer electrons of every atom involved in bonding with other atoms. Thermal shaking of the atoms results in some of the bonds breaking and freeing electrons. When a silicon atom looses an electron, we can consider there to be a hole in its valence electrons (Figure 5.49(a)). When electrons are made to move as a result of the application of a potential difference, i.e. an electric field, they can be thought of as hopping from valence site into a hole in a neighbouring atom, then to another hole, etc. Not only do electrons move through the material but so do the holes, the holes moving in the opposite direction to the electrons. We can think of the above behaviour in the way shown in Figure 5.49(b). One way of picturing this behaviour is in terms of a queue of people at, say, a bus-stop. When the first person gets on the bus, a hole appears in the queue between the first and second person. Then the second person moves into the hole, which now moves to between the second and third person. Thus as people move up the queue, the hole moves down the queue.

The conductivity of a semiconductor can be very markedly changed by impurities. For this reason the purity of semiconductors must be very carefully controlled. With the silicon used for the manufacture of semiconductor devices, the impurity level is routinely controlled to less than one atom in a thousand million silicon atoms. Foreign atoms can, however, be deliberately introduced in controlled amounts into a semiconductor in order to change its electrical properties. This is referred to as *doping*. Atoms such as phosphorus, arsenic or antimony when added to silicon add easily-released electrons and so make more electrons available for conduction. Such dopants are called *donors*. Semiconductors with more electrons available for conduction than holes are called an *n-type semiconductor*. Atoms such as boron, gallium, indium or aluminium add holes into which electrons can move. They are thus referred to as *acceptors*. Semiconductors with an excess of holes are called a *p-type semiconductor*.

Activity

Thermistors are temperature sensors made from mixtures of metal oxides, e.g. the oxides of chromium, cobalt, iron, manganese and nickel. These oxides are semiconductors. Determine how the resistance of a thermistor changes as the temperature changes from room temperature to 100°C.

Problems

1 Explain the term *grain* when used in connection with the structure of metals.
2 Explain what is meant by the term *alloy*.
3 Explain the terms *ferrous alloy* and *non-ferrous alloy*.
4 Describe the structure of metals.
5 How does the grain size in a metal affect its properties?
6 How does the shape of grains within a metal affect its properties?
7 Describe the effects on the grain structure and properties of a metal of cold working.
8 Describe the effects on the properties of carbon-steels of increasing the percentage of carbon on the alloy.
9 What types of structure might you expect for a metal which is (a) ductile, (b) brittle?
10 A pure metal is formed into an alloy by larger atoms being forced into the spaces in its crystal structure. What changes might be expected in the properties and why?
11 Describe the difference between amorphous and crystalline polymer structures and explain how the amount of crystallinity affects the mechanical properties of the polymer.
12 Compare the properties of low- and high-density polyethylene and explain the differences in terms of structural differences between the two forms.
13 Why are (a) stabilisers, (b) plasticisers and (c) fillers added to polymers?
14 Describe how the properties of PVC depend on the amount of plasticiser present in the plastic?
15 Increasing the amount of sulphur in a rubber increases the amount of cross-linking between the molecular chains. How does this change the properties of the rubber?
16 Explain how elastomers can be stretched to several times their length and still be elastic and return to their original length.
17 Calculate the tensile modulus of a composite consisting of 45% by volume of long aligned glass fibres, tensile modulus 76 GPa, in a polyester matrix, tensile modulus 4 GPa. In what direction does your answer give the modulus?
18 In place of the glass fibres referred to in problem 17, carbon fibres are used. What would be the tensile modulus of the composite if the carbon fibres had a tensile modulus of 400 GPa?
19 Long boron fibres, tensile modulus 340 GPa, are used to make a composite with aluminium as the matrix, the aluminium having a tensile modulus of 70 GPa. What would be the tensile modulus of the composite in the direction of the aligned fibres if they constitute 50% of the volume of the composite?

20 How will the properties of composites differ if they are (a) made of long fibres all orientated in the same direction, (b) short fibres with random orientation?

21 State what structural changes take place, and the consequential changes in properties, in (a) annealing, (b) quenching, (c) tempering, (d) precipitation hardening, (e) flame hardening, (f) case hardening.

22 A carbon steel is found to have the following properties. Explain how they arise in terms of the structure of the steel.

	Strength MPa	Hardness HB	Elongation %
As rolled	550	180	32
Annealed	465	125	32
Quenched, tempered 200°C	850	495	17
Quenched, tempered 650°C	585	210	32

23 An aluminium-magnesium-silicon alloy is found to have the following properties. Explain how they arise in terms of the structure of the alloy.

	Strength MPa	Hardness HB	Elongation %
Annealed	125	30	25
Precipitation hardened	310	95	12

24 The striking part of the head of a hammer is required to be very hard, but the main body of the hammer head is required to be softer and more tough. How can these properties be achieved in a single piece of steel?

25 What properties are required of a hacksaw blade and how might they be achieved?

6 Processing of materials

6.1 Shaping metals

The main methods used to shape metals are:

1 *Casting*
A product is formed by pouring liquid metal into a mould. Sand casting involves using a mould made of sand, die casting a metal mould.

2 *Manipulative processes*
A shape is produced by plastic deformation processes. This includes such cold working methods as rolling, drawing, pressing and impact extrusion. Hot working processes include rolling, forging and extrusion.

3 *Powder techniques*
A shape is produced by compacting a powder.

4 *Material removal*
This involves producing a shape by metal removal, e.g. machining.

The above shaping processes are one way of producing a product. Another way is metal joining, of which the main processes are:

1 *Chemical methods*
Sticking a stamp on an envelope is an example of using a chemical joining method, i.e. an adhesive, to join a stamp to an envelope.

2 *Physical methods*
Soldering a wire to an electrical contact is an example of such a method. In this group of methods we can include soldering, brazing and welding. These methods depend on changes of state from liquid metal to solid metal to make the joint.

3 *Mechanical methods*
These involve fastening systems, e.g. rivets, bolts and nuts, and they depend on stresses set up by the fastener to hold items together.

The shaping or assembly method used for a particular product will depend on the metal to be used, its form of supply, and the form of the product. This chapter discusses the commonly used shaping methods, any structural changes that occur in materials as a result of such processes and the relationship between the properties of materials and the processing methods that can be used.

6.2 Casting

Casting is the shaping of an object by pouring the liquid metal into a mould and then allowing it to solidify and form a product with the internal shape of the mould. The resulting shape may be that of the final manufactured object, or one that requires some machining, or even an ingot which is then further processed by manipulative processes.

The metal has to flow into all parts of the mould. Think of the problems of pouring treacle into a mould compared with pouring water. With the water it is much easier to fill all parts of the mould. Thus alloys used for casting where gravity is used to get the liquid metal to flow, as in sand casting and gravity die casting, have their alloying constituents chosen to give good flowing properties, i.e. low viscosity. Where pressure is used to force the liquid metal into the mould, as in pressure die casting, then a more viscous alloy can be used.

The grain structure within the product is determined by the rate of cooling (Figure 6.1). Where the cooling rate is high then small grains are produced, where the rate is low then larger grains are produced. Thus, since the metal in contact with the mould cools faster than that in the centre of the casting, smaller grains called *chill crystals* are produced at the mould surfaces than in the centre of a casting. The cooling rate a little in from the mould walls is less than that at the walls. A consequence of this is that some chill crystals, given time, can develop into long elongated crystals in an inward direction; these are called *columnar crystals*. The grains produced in the centre of the casting where the cooling rate is the slowest are called *equiaxed crystals*. These crystals grow in liquid metals which is constantly on the move due to convection currents. As a consequence, the crystals are almost spherical.

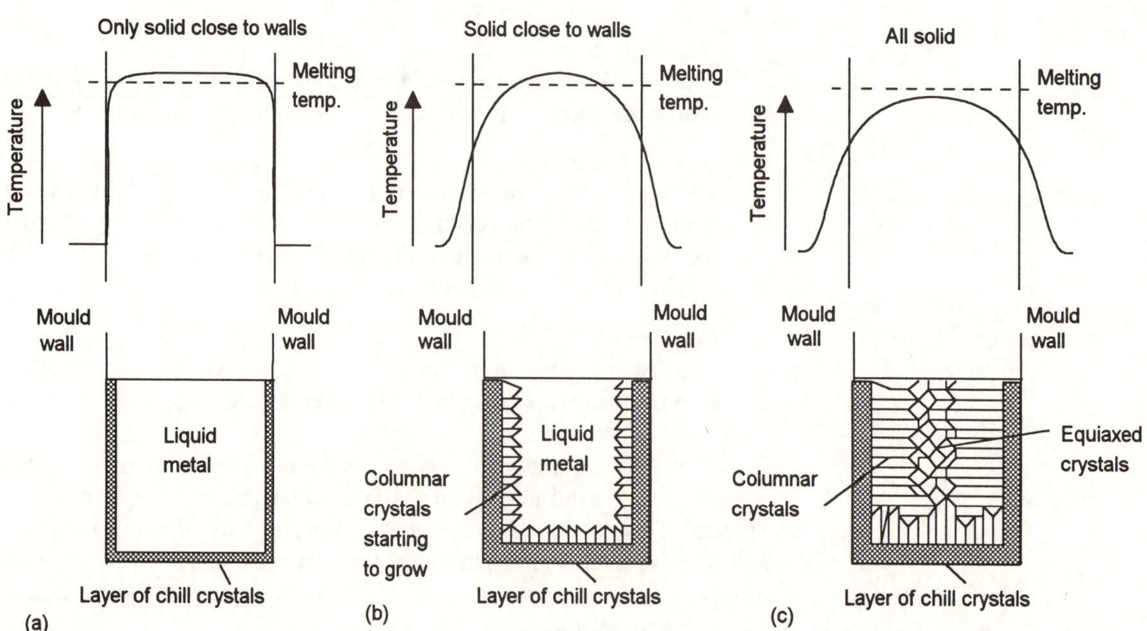

Figure 6.1 *Stages in the solidification of a casting: (a) initial cooling, (b) after further cooling, (c) when completely solid*

In general, a casting structure having entirely small equiaxed crystals is preferred. This type of structure can be promoted by a more rapid rate of cooling for the casting. Castings in which the mould is made of sand tend to have a slower rate of cooling as sand has a low thermal conductivity. Thus sand castings tend to have large columnar grains and hence relatively low strength. Die casting involving metal moulds has a much faster rate of cooling and so gives castings having a bigger zone of equiaxed crystals. As these are smaller than columnar crystals, the casting has better properties. Table 6.1 shows the types of differences that can occur with aluminium casting alloys.

Table 6.1 *Effect of process on properties of aluminium alloy castings*

Aluminium alloy	Tensile strength MPa		Percentage elongation	
	Sand cast	Die cast	Sand cast	Die cast
5% Si, 3% Cu	140	150	2	2
12% Si	160	185	5	7

Castings do not show directionality of properties, the properties being the same in all directions. They do, however, have the problems produced by blowholes and other voids occurring during the solidification, e.g. from trapped air, the evolution of gas from the liquid metal as it cools and from local solidification shrinkage.

Example
What type of grain structure might be expected when liquid metal is poured into a narrow metal mould?

The rate of cooling will be high because it is a metal mould. Also, because the mould is thin, all parts of the casting will cool quickly. Thus it is likely that chill crystals will be formed throughout since there is not enough time for columnar crystals to develop.

6.2.1 Casting methods

There are a number of casting methods possible and the factors determining the choice of a particular method are:

1. Size, complexity and dimensional accuracy required
2. The number of castings required
3. The cycle time, i.e. the time taken to complete a casting and then be ready to repeat the process
4. The flexibility of the casting process to be adaptable to different forms of product
5. The operating cost per casting
6. The mechanical properties required of the casting

7 The quality of the casting, i.e. surface finish, porosity, non-metallic inclusions

Commonly used casting methods are sand casting and die casting.

Sand casting involves the making of a mould using a mixture of sand with clay. This is packed around a pattern of the casting, generally of a hard wood and larger than the required casting to allow for shrinkage. The mould is made in two or more parts so that the pattern can be extracted after the sand has been packed round it. Molten metal is poured into the cavity, risers allow for surplus molten metal to compensate for contraction of the metal as it cools. The mould is broken to extract the product. Thus each mould is destroyed after making just one casting, though the pattern which was used to make the mould can be used repeatedly to make further moulds. Figure 6.2 shows a sectional view of an example of a mould.

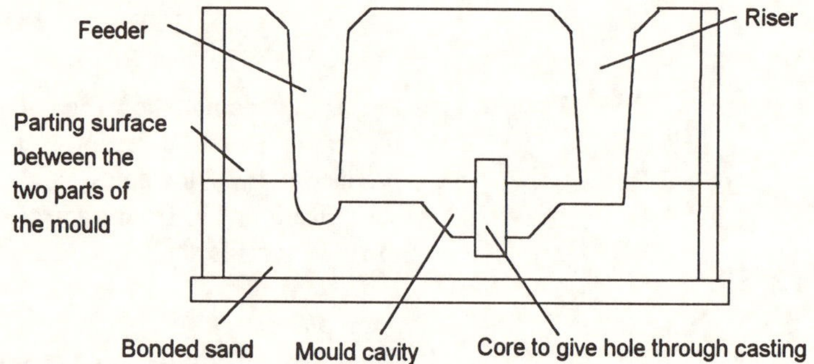

Figure 6.2 *Sectional view of a mould*

Sand casting can be used for most metals and for a wide range of casting sizes. Patterns are relatively cheap to make and mould making is also relatively easy. It is thus often the cheapest process for small-number production. It does, however, have a long cycle time since the rate at which heat transfers out of the casting is slow and thus quite some time can elapse after pouring the metal into the mould before the casting is ready. Because of the low thermal conductivity of the mould material, large columnar crystals occur and thus the mechanical properties tend to be poor. Surface finish is also poor and porosity and non-metallic inclusions in the casting tend to be common. It is thus not suitable for products requiring close specification of tolerances without further processing. Typical products are engine blocks, machine tool bases, pump housings and cylinder heads.

With *die casting* the same mould is used for a large number of castings. Such moulds have to be made of a material which will withstand the temperature changes and wear associated with repeated castings and thus are made of metal. This tends to restrict the metals that can be cast to relatively low melting points, e.g. light alloys and some steels and cast irons. Two types of die casting are used, gravity and pressure die casting.

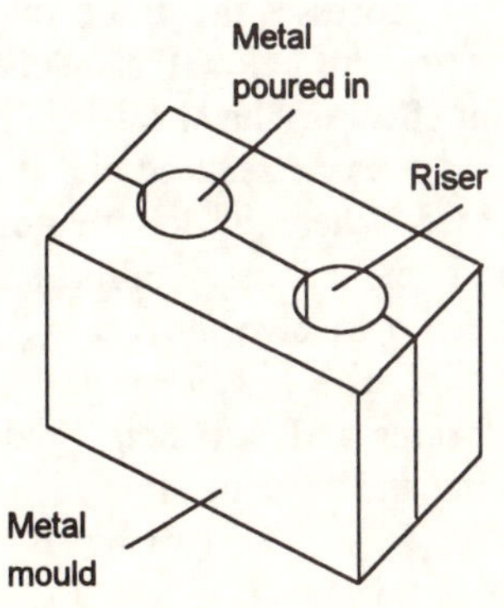

Figure 6.3 *Gravity die casting*

Gravity die casting has the molten metal poured into the mould; the head of liquid metal is responsible for forcing the metal to flow into the various parts of the mould (Figure 6.3). This method is mainly used for small, simple shapes with only the use of simple cores for items such as holes; surface texture is good. The cycle time is limited by the rate of heat transfer out of the casting and, since the mould is metal, this is faster than occurs with sand casting. The metal has to flow quickly to all parts of the mould before solidification occurs and, since the metal solidifies quickly as a result of being in contact with metals walls which have high thermal conductivity, usually just low melting point alloys such as those of aluminium, copper, magnesium and zinc can be used. It is commonly used for casting less that 5 kg in weight. Typical products are cylinder heads, pistons and connecting rods.

With *pressure die casting* the liquid metal is injected into a water-cooled mould under pressure (Figure 6.4). This has the advantage that the metal can be forced into all parts of the mould cavity and thus very complex shapes with high dimensional accuracy can be produced. The method is limited to low melting point alloys, mainly those of zinc and aluminium. Typical products are in the size range 10 g to 50 kg and include engine parts, toy parts, pump components and domestic appliance parts.

With die casting, the cost of the mould is high and thus the process is relatively uneconomic for small-number production; large-number production is necessary to spread the cost of the mould. These initial high costs may, however, be more than compensated for by the high quality of the surfaces and hence the reduction or elimination of machining or finishing costs.

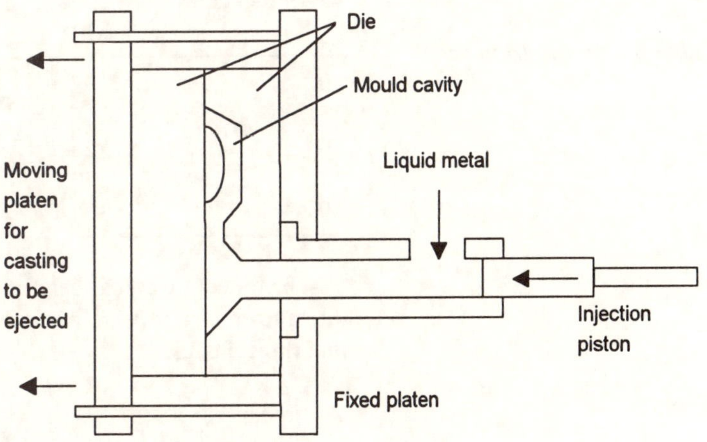

Figure 6.4 *Pressure die casting*

6.3 Manipulative processes

Manipulative processes involve the shaping of a material by plastic deformation and require the material to be ductile and not brittle. Figure 6.5 shows the basic form of the stress–strain graph for a ductile material. Plastic deformation starts to occur when the stress exceeds the yield

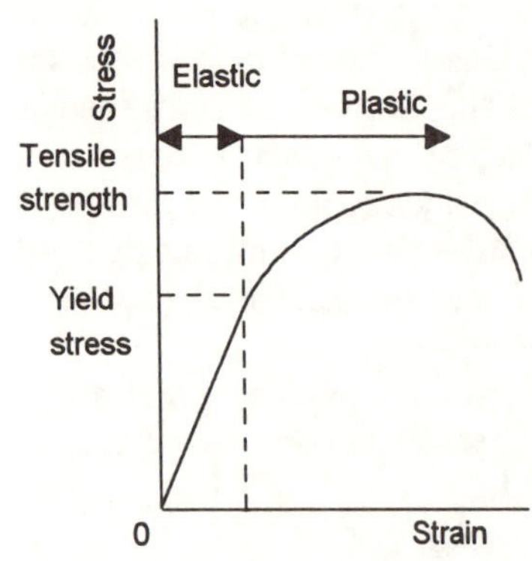

Figure 6.5 *Stress–strain graph*

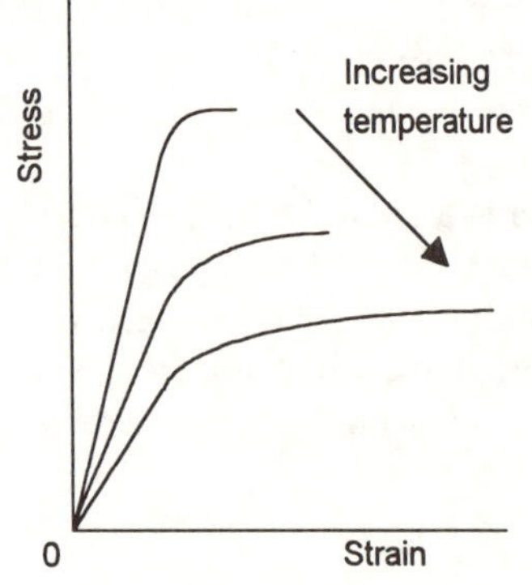

Figure 6.6 *Stress–strain graph*

stress so the forces required for manipulative processes are those that give rise to stresses in excess of the yield stress, but below that of the tensile strength. Figure 6.6 shows the general effect on the stress–strain graph of increasing the temperature. The yield stress is reduced and the amount of plastic deformation possible increased. Where the deformation is carried out at a temperature in excess of the recrystallisation temperature of the metal, the process is said to involve *hot working*; below the recrystallisation temperature is it called *cold working*. Hot-working processes include hot rolling, forging and extrusion. Cold-working processes include cold rolling, drawing and pressing.

6.3.1 Working

During cold working, the crystal structure becomes broken up and distorted, leading to an increase in mechanical strength and hardness and a decrease in ductility, this being termed *work hardening*. The more the material is worked the harder and more brittle the material becomes. Table 6.2 shows the effect on the mechanical properties of work hardening when a sheet of annealed aluminium is cold rolled and its thickness reduced. A stage in the working can be reached when the material becomes too hard and too brittle to be further worked. With the rolled aluminium sheet referred to in Table 6.2, this condition has been reached with about a 60% reduction in sheet thickness. The material is then said to be *fully work hardened*.

Table 6.2 *Effect of work hardening on mechanical properties*

% reduction in sheet thickness	Tensile strength MPa	Percentage elongation	Hardness HV
0	92	40	20
15	107	15	28
30	125	8	33
40	140	5	38
60	155	3	43

When a cold-worked metal is heated the events that occur depend on the temperature to which it is heated. The events can be broken down into three phases:

1 *Recovery*
When a cold-worked metal is heated to temperatures up to about $0.3T_m$, where T_m is the melting point on the Kelvin temperature scale of the metal concerned, then the internal stresses resulting from the working start to become relieved. There are no changes in grain structure during this but just some slight rearrangement of atoms in order that the stresses become relieved. This process is known as *recovery*. Copper has a melting point of 1083°C, or 1356 K. Hence stress relief with copper requires heating to about 407 K, i.e. 134°C.

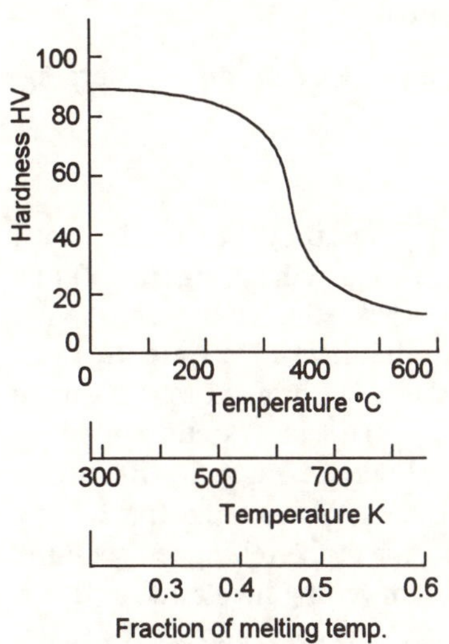

Figure 6.7 *The effect of heat treatment on cold-worked copper*

2 *Recrystallisation*

If the heating is continued to a temperature of about 0.3 to $0.5T_m$ there is a very large decrease in hardness, decrease in strength and increase in elongation. Figure 6.7 shows the effect on the hardness. The grain structure of the metal changes, the metal recrystallising. With *recrystallisation*, crystals begin to grow from nuclei in the most heavily deformed parts of the metal. The onset of recrystallisation is about 150°C for aluminium, 200°C for copper, 450°C for iron and 620°C for nickel.

3 *Grain growth*

As the temperature is further increased from the recrystallisation temperature, so the crystals grow until they have completely replaced the original distorted cold-worked structure. The hardness, tensile strength and percentage elongation change little during this phase, the only change being that the grains grow.

Figure 6.8 summarises the above effects.

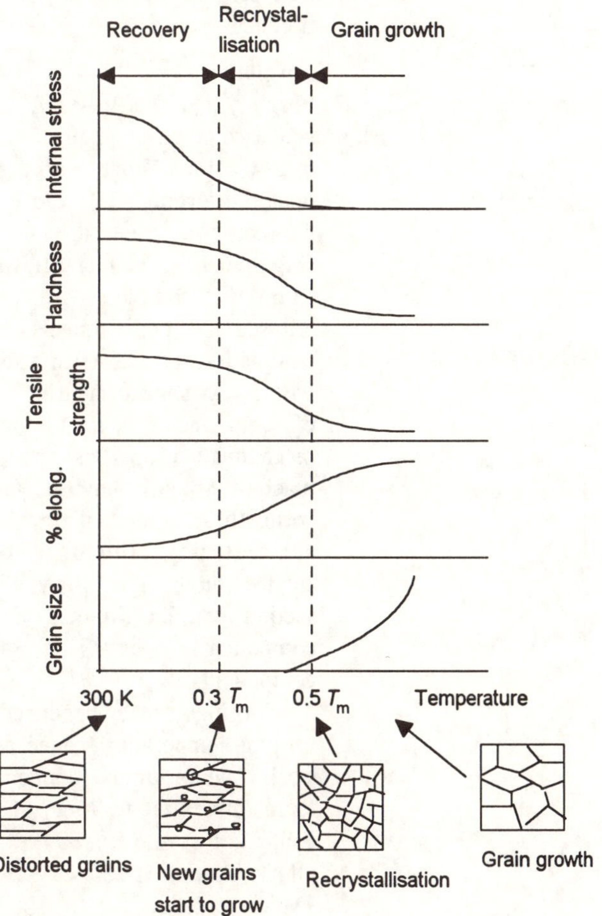

Figure 6.8 *The effect of an increase in temperature on cold-worked materials*

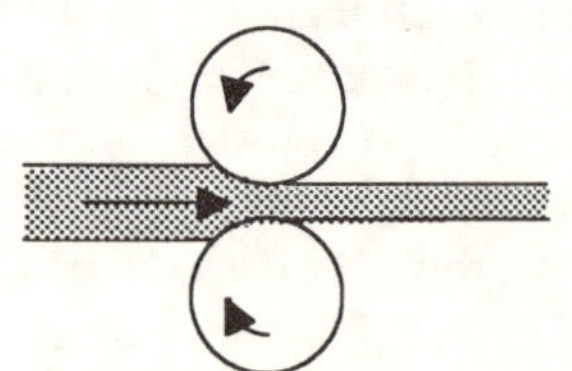

Figure 6.9 *Rolling*

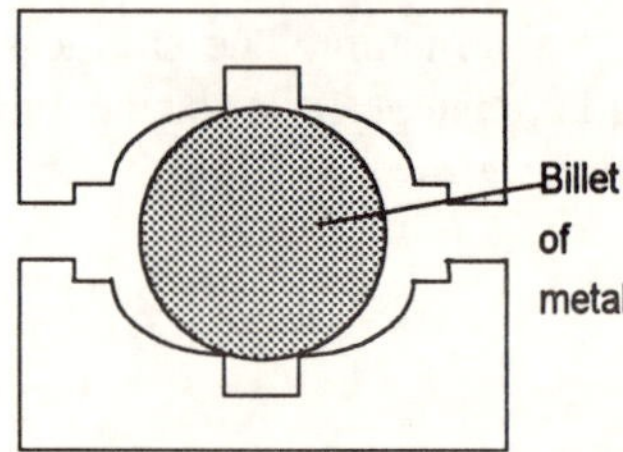

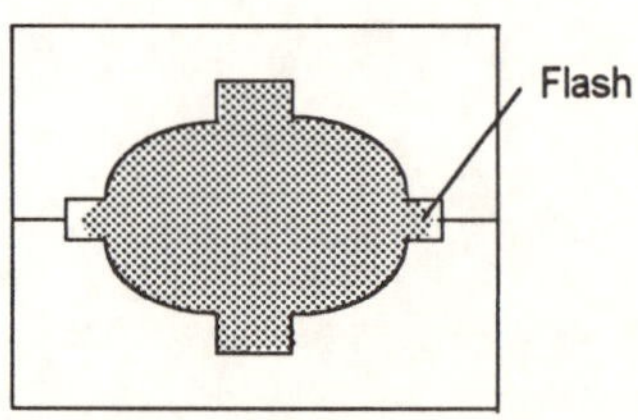

Figure 6.10 *Closed die forging*

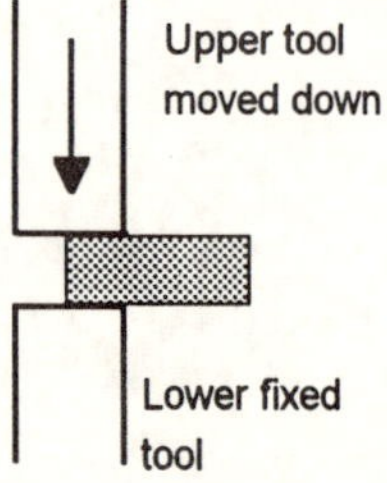

Figure 6.11 *Open die forging*

6.3.2 Commonly used manipulative methods

Manipulative methods commonly used include rolling, forging, extrusion, drawing and sheet forming.

1 *Rolling*

Rolling is the continuous shaping of a metal by passing the metal between the gap between a pair of rotating rollers (Figure 6.9) and can be a hot- or cold-working process depending on the temperature of the material being rolled. When cylindrical rollers are used, the product is in the form of a bar or sheet, but profiled rollers can be used to produce contoured surfaces, e.g. structural sections used for window frames with the rolled product only having to be trimmed to size and joined with other rolled shapes to make the frame. Any metal can be rolled, provided it is ductile at the rolling temperature. Hot rolling is usually the first step in converting ingots and billets to the required shape. Sheet and strip are often cold rolled as a cleaner, smoother finish to the metal surfaces is produced than hot rolling. The process also gives, as a result of work hardening, a harder product.

2 *Forging*

Forging is a hot-working process and involves the metal being squeezed between a pair of dies, the metal having to be ductile at the forging temperature. Forging can be either closed die forging or open die forging. With *closed die forging* the hot metal is squeezed between two shaped dies which effectively form a complete mould (Figure 6.10). The metal flows under the pressure into the die cavity. In order to completely fill the die cavity, a small excess of metal is allowed and this is squeezed outwards to form a flash which is later trimmed away. Closed die forging can be automated and can thus have a very short cycle time. It can be used to produce large numbers of components with high dimensional accuracy and better mechanical properties than would be produced by casting or machining. With *open die forging* the hot metal is hammered by a vertically moving tool trapping the metal against a stationary tool and squeezing it (Figure 6.11). This type of forging is like that once carried out by the village blacksmith. By repeated hammering a section can be thinned. Closed die forging is used for small components, open die forging normally being used for larger components.

Forgings can be produced from any metal that is ductile at the forging temperature concerned; the method is mainly used with steels, aluminium, copper and magnesium alloys. Forging temperatures range from 920 to 1370°C for steels, 590 to 920°C for copper alloys and from 340 to 480°C for aluminium and magnesium alloys. Product sizes vary from about 10 g to 250 kg in weight. Typical products are tool bodies, airframe components, axle shafts and connecting rods.

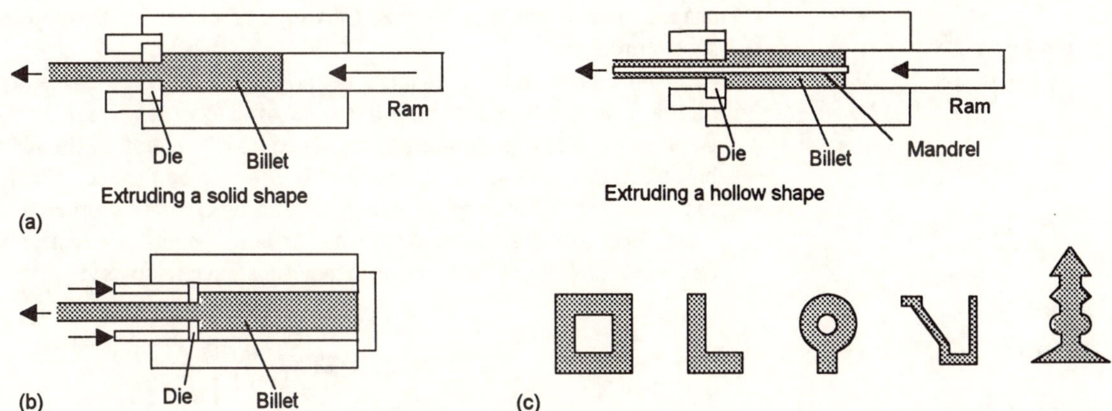

Figure 6.12 *(a) Direct extrusion, (b) indirect extrusion, (c) examples of sections*

3 *Extrusion*

Extrusion involves metal being forced, under pressure, to flow through a die, i.e. a shaped orifice, to give a product of a small cross-sectional area. The materials used must be ductile and this can mean hot extrusion, the higher temperature often being needed to reduce to manageable levels the pressure required; typically 350–500°C for aluminium alloys, 700–800°C for brasses and 1100–1250°C for steels.

With *direct extrusion* (Figure 6.12(a)) a heated billet is forced through a die. It is rather like squeezing toothpaste out of its tube. Long lengths of quite complex section, including hollow sections, can be extruded. With *indirect extrusion* (Figure 6.12(b)) a heated billet of metal is extruded through a die by the die being pushed into the billet. A wide variety of sections can be produced by hot extrusion, many of which would be uneconomic to produce by any other method. Figure 6.12(c) shows some examples.

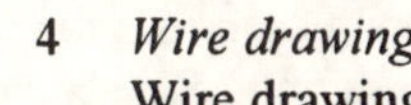

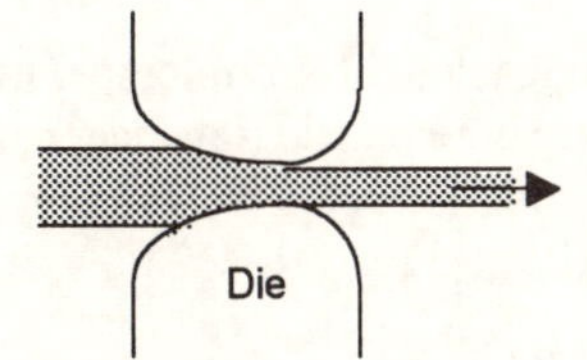

Figure 6.13 *Wire drawing*

4 *Wire drawing*

Wire drawing involves the pulling in of metal through a die (Figure 6.13) in order to reduce the diameter to that required. A number of stages are often used as cold working hardens the metal and so there may be annealing operations between the various drawing stages in order to soften the material so that further drawing can take place.

5 *Sheet formation*

The term *deep drawing* is used for the forming of sheets in which a sheet metal blank is pushed into a die aperture by a punch (Figure 6.14), the action causing the metal to flow into the die. The blank is not clamped round its rim and so the sheet is pulled into the die by the action. The term deep drawing tends to be just used for when the depth of the product is one or more times its diameter and the term *stamping* is often used if it is less, though in practice there is no real difference in the operation. The more ductile materials such as aluminium, brass and mild steel are used and the products are

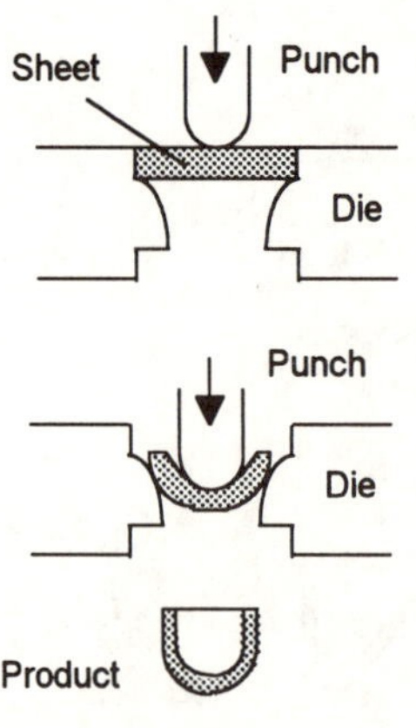

Figure 6.14 *Deep drawing*

typically deep-recessed parts with vertical walls, e.g. cup- and box-shaped articles, without flanges. Cartridge cases are an example of such a product.

Stretch forming involves sheet being gripped round its edges and then a form block pushed into the sheet to deform it (Figure 6.15). Because the blank is clamped round its edges, the action of the form block causes the sheet to be stretched in order to be pushed into the required shape. The term *pressing* is often used for this process, it being used to form such products as car body panels, kitchen pans and other cooking utensils. Ductile materials have to be used.

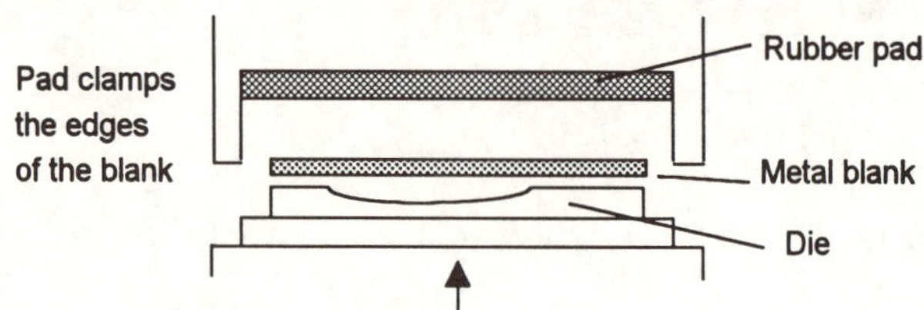

Figure 6.15 *Stretch drawing*

Activity

The bodywork panels of cars are formed by pressing. Suggest the form the die might take for a door panel?

6.4 Powder processing

Powder processing involves (Figure 6.16):

1 *Blending the metal powder to obtain the desired properties*
Iron, copper and graphite particles might be blended so that when sintered the required alloy is produced. A lubricant might also be added to enable the particles to flow more easily over each other and so pack better together, giving a higher density product.

2 *Compacting the powder*
The powder is placed in a die and compacted, at room temperature, to form the required shape component. Some plastic flow occurs as a result of this pressure and the component, termed a *green compact*, retains its shape when removed from the die.

3 *Sintering to produce the solid component*
The green compact is then heated to a high temperature so that the particles coalesce to form the solid component. If the compacted powder had been heated in the die there would have been the problem of having a die material which could withstand the high temperature. If a lubricant is used, then this would be burnt off.

Powder processing eliminates the need for any machining since the dimensional accuracy and surface finish obtained is good. It is a useful method for the production of components which would be difficult to cast because of high melting point alloys or difficult to manipulate because of being brittle. The shapes produced tend to be simple and relatively small.

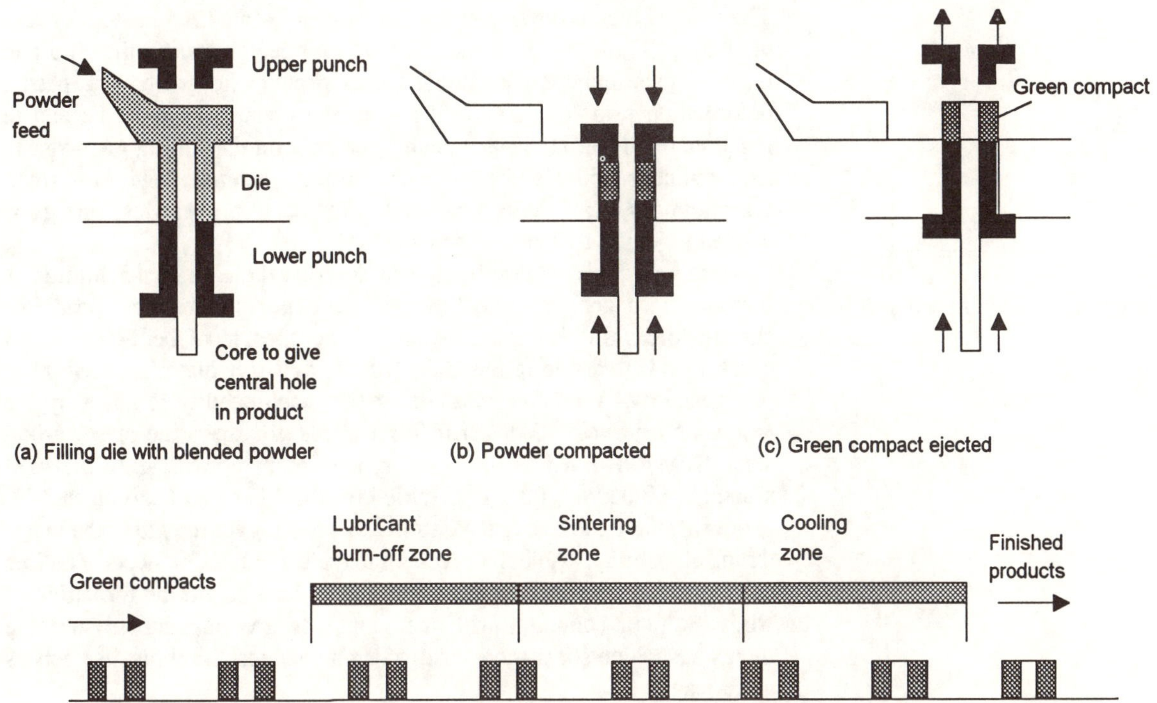

Figure 6.16 *Compacting and sintering*

Typical applications include iron–8% carbon for moderately loaded gears, iron–5–20% copper–0.8% carbon for medium loaded gears, nickel alloy steel 4% nickel–1% copper–0.7% carbon for wear resisting, high stress components such as differential and transmission gears, copper–nickel–zinc for corrosion resisting conditions such as gears for use in marine environments.

6.5 Material removal

Material removal methods involve the shaping of products by the selective removal of material. Such methods are generally a secondary process, following a primary process such as casting or forging, and are used to produce the final shape to the required accuracy and surface finish. Invariably waste material is produced and thus any costing of a material removal process has to allow for this.

Cutting is a method of material removal which create chips; this including single-point cutting methods such as planing and turning and multi-point cutting methods such as drilling, sawing and filing, and those methods which do not create chips such as blanking and shearing. In cutting, the tool causes the workpiece material at the cutting edge to become highly stressed and subject to plastic deformation. The more ductile the material the greater the amount of plastic deformation and the

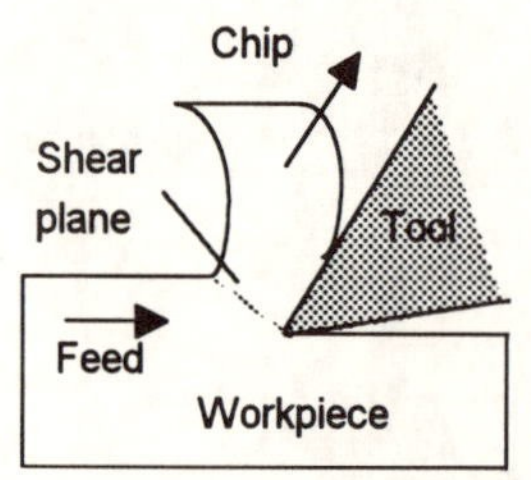

Figure 6.17 *Chip formation*

more the material of the workpiece spreads along the face of the tool. The deformed chip flows over the tool surface, generating heat as a result of friction (Figure 6.17). It is difficult to lubricate between the chip and the tool face and high frictional forces occur. The greater the plastic deformation prior to fracture of a chip the greater the force needed to machine the material. Thus a ductile material on machining gives rise to a continuous chip while a more brittle material leads to small discontinuous chips being produced with the result that less energy is needed for cutting when this happens.

The term *machinability* is used to describe the ease of machining. A material with good machinability will produce small chips, need low cutting forces and energy expenditure, be capable of being machined quickly and give a long tool life. Ductile and soft materials have poor machinability. A relative measure of the machinability is given by the *machinability index*. With British Standards, for steels the plain carbon steel 070M20 is rated as 100% and others compared with it. In the AISI system, an index of 100% is specified for the 1212 plain carbon steel. It is only a rough guide to machinability, but the higher the index the better the machinability. Table 6.3 gives some typical values for steels. The free cutting steels have additives, e.g. sulphur or lead, to aid the formation of chips. Sulphur combines with manganese to give manganese sulphide inclusions which shear more easily. The manganese sulphide also acts as a lubricant.

Table 6.3 *Machinability*

Material	Machinability index
Plain carbon steel	
070M20, 0.20% carbon	100
080M30, 0.30% carbon	70
080M40, 0.40% carbon	70
080M50, 0.50% carbon	50
070M55, 0.55% carbon	50
Free-cutting steels	
210M15	200
214M15	140
220M07	200

6.6 Forming processes with polymers

Many polymer-forming processes are essentially two stage, the first stage being the production of the polymer in a powder, granule or sheet form and the second stage being the shaping of this material into the required shape. The first stage can involve the mixing with the polymer of suitable additives and other polymers in order that the finished material should have the required properties. Second-stage processes for thermoplastics generally involve heating the powder, granule or sheet material until it softens, shaping the softened material to the required shape and then cooling it. For thermosets the second-stage processes involve forming the thermosetting materials to the required shape and then heating them so that they undergo a chemical change to cross-link polymer chains into a highly linked polymer. The main second-stage processes used for forming polymers are:

1 *Moulding*
  This includes injection moulding, reaction injection moulding, compression moulding and transfer moulding.
2 *Forming*
  This includes such processes as extrusion, vacuum forming, blow moulding and calendering.
3 *Cutting*

In addition, products may be formed by polymer joining. The main processes are:

1 Adhesives
2 Welding
3 Fastening systems such as riveting, press and snap fits and screws

The choice of process will depend on a number of factors, such as:

1 The quantity of items required
2 The size of the items
3 The rate at which the items are to be produced, i.e. cycle time
4 The requirements for holes, inserts, enclosed volumes, threads
5 Whether the material is thermoplastic or thermoset

6.7 Moulding

Moulding uses a hollow mould to form the product. The main processes are injection moulding, reaction injection moulding, compression moulding and transfer moulding.

6.7.1 Injection moulding

A widely used process for thermoplastics, though it can also used for rubbers, thermosets and composites, is *injection moulding*. With this process, the polymer raw material is pushed into a cylinder by a screw or plunger, heated and then pushed, i.e. injected, into the cold metal mould (Figure 6.18). The pressure on the material in the mould is maintained while it cools and sets. The mould is then opened and the component extracted, and then the entire process repeats itself. The sequence is thus:

1 Mould closed
2 Nozzle forward

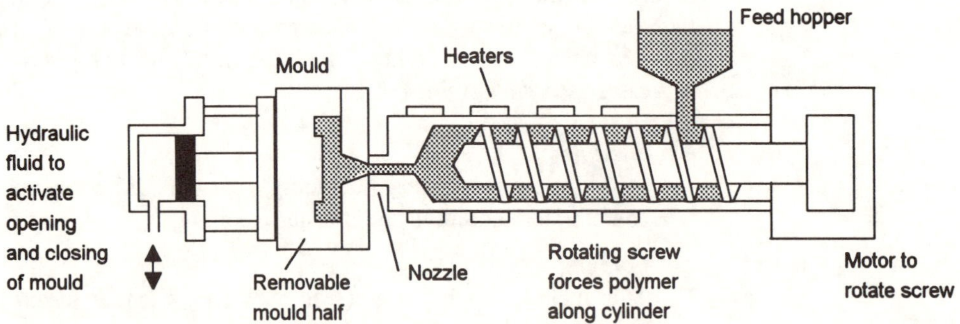

Figure 6.18 *Injection moulding*

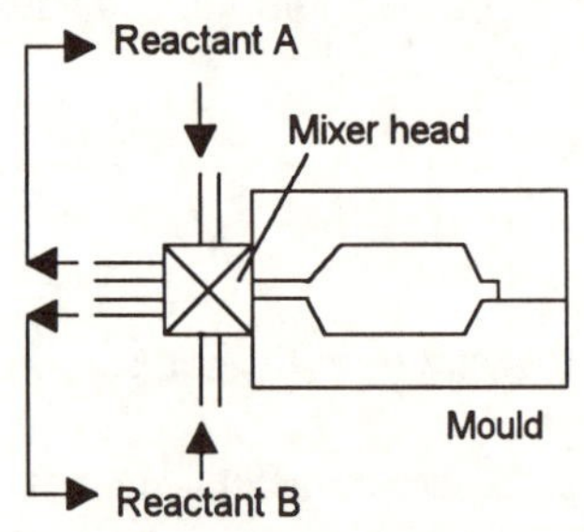

Figure 6.19 *Reaction injection moulding*

3 Screw forward
4 Hold injection pressure
5 Screw rotates and retracts
6 Nozzle retracts
7 Mould opens and moulding ejected
8 Procedure then repeated

High production rates can be achieved and complex shapes with inserts, threads, holes, etc. produced; sizes range from about 10 g to 25 kg in weight. Typical products are beer or milk bottle crates, toys, control knobs for electronic equipment, tool handles, pipe fittings.

6.7.2 Reaction injection moulding

Reaction injection moulding involves the reactants being combined in the mould to react and produce the polymer (Figure 6.19). The choice of materials that are processed in this way is determined by the reaction time, this must be short, e.g. 30 s, so that cycle times are short. It is mainly used with polyurethanes, polyamides and polypropylene oxide and composites incorporating glass fibres. The preheated reactants are injected at high speed into a closed mould where they fill the mould and combine to produce the finished product. This method is used for large automotive parts such as spoilers, bumpers and front and rear fascia.

Activity
Design a mould for a car bumper.

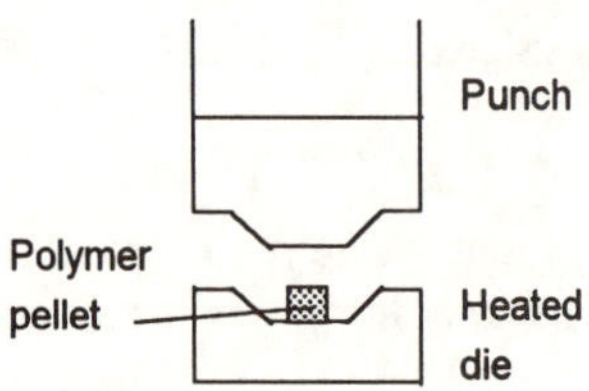

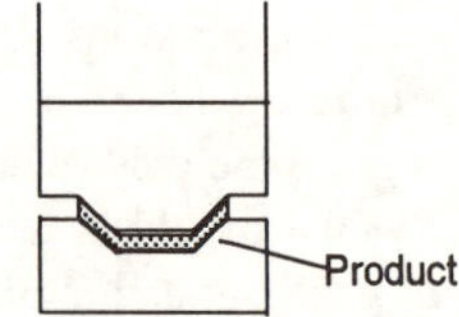

Figure 6.20 *Compression moulding*

6.7.3 Compression moulding

Compression moulding is widely used for thermosets. The powdered polymer is compressed between the two parts of the mould and heated under pressure (Figure 6.20) to initiate the polymerisation reaction. The process is limited to relatively simple shapes from a 2–3 g to 15 kg in weight. Typical products are dishes, handles and electrical fittings.

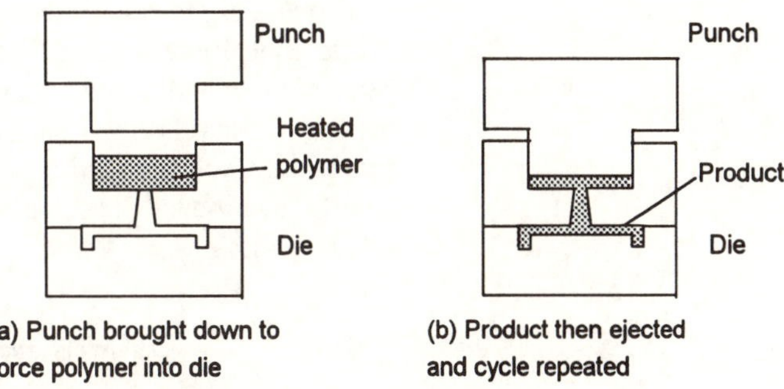

Figure 6.21 *Transfer moulding*

Transfer *moulding* differs from compression moulding in that the powdered polymer is heated in a chamber before being transferred by a plunger into the heated mould (Figure 6.21).

6.8 Forming processes

Forming processes involve the flow of a polymer through a die to form the required shape.

6.8.1 Extrusion

A very wide variety of plastic products are made from extruded sections, e.g. curtain rails, household guttering, window frames, polythene bags and film. *Extrusion* (Figure 6.22) involves the forcing of the molten thermoplastic polymer through a die. The polymer is fed into a screw mechanism which takes the polymer through the heated zone and forces it out through the die. In the case of an extruded product such as curtain rail, the extruded material is just cooled.

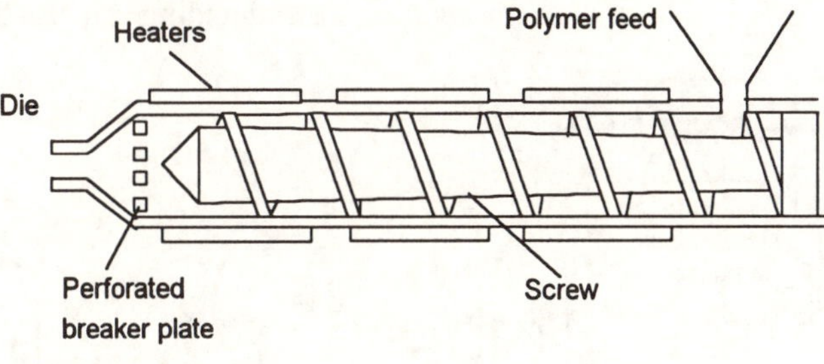

Figure 6.22 *Extrusion*

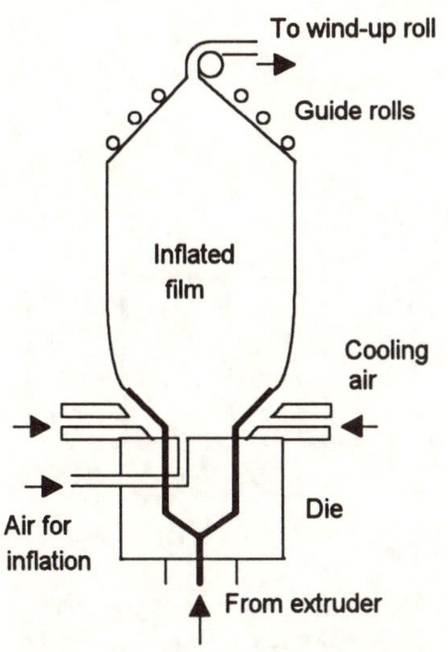

Figure 6.23 *Blown film extrusion*

If thin film or sheet is required, a die may be used which gives an extruded cylinder of material. This cylindrical extruded material is inflated by compressed air while still hot to give a tubular sleeve of thin film (Figure 6.23). The expansion of the material is accompanied by a reduction in thickness. Such film can readily be converted into bags. Polyethylene is readily processed to give tubular sleeves by this method but polypropylene presents a problem in that the rate of cooling is inadequate to prevent crystallisation and so the film is opaque and rather brittle. Flat film extrusion (Figure 6.24) can be produced using a slit-die. The rate of cooling, by the use of rollers, can be made fast enough to prevent crystallisation occurring with polypropylene. The extrusion process can be used with most thermoplastics and yields continuous lengths of product. Intricate shapes can be produced and a high output rate is possible.

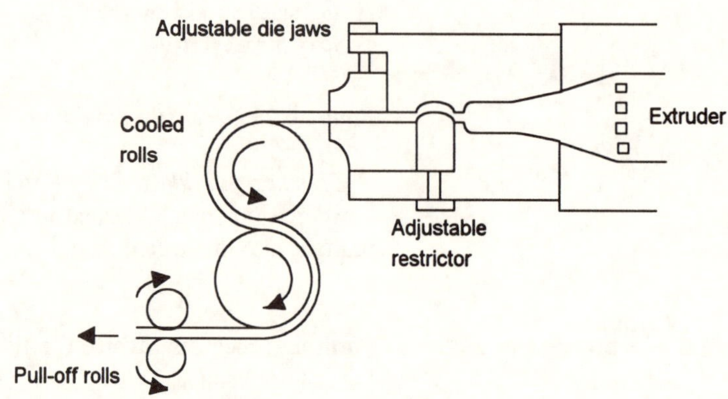

Figure 6.24 *Flat film extrusion*

6.8.2 Blow moulding

Blow moulding is a process used widely for the production of hollow articles such as plastic bottles from thermoplastics. Containers from as small as 10^{-6} m^3 to as large as 2 m^3 can be produced. With *extrusion blow* moulding the process involves the extrusion of a hollow thick-walled tube which is then clamped in a mould (Figure 6.25). Pressure is applied to the inside of the tube to inflate it so that it fills the mould. Blow moulding can also be used with injection moulding.

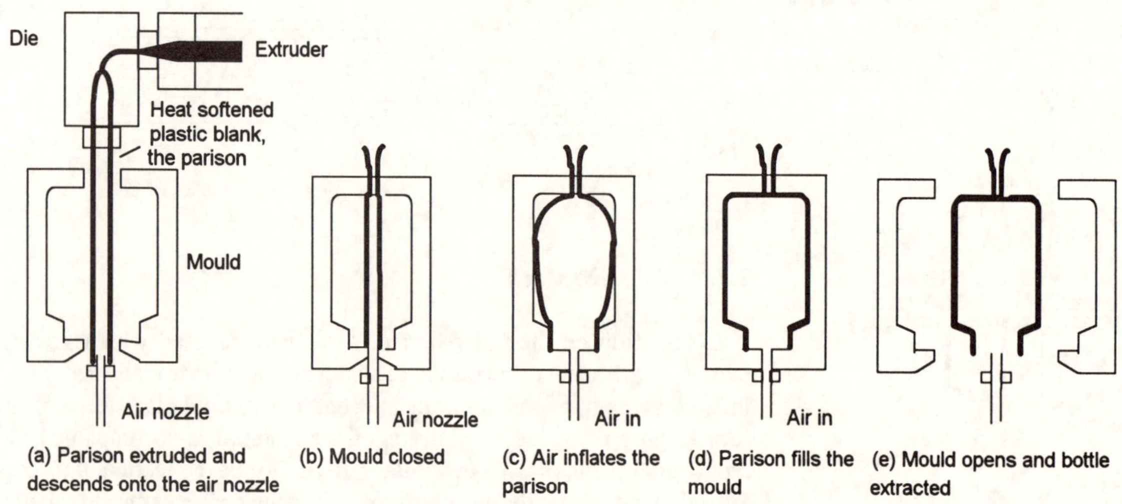

Figure 6.25 *Blow moulding*

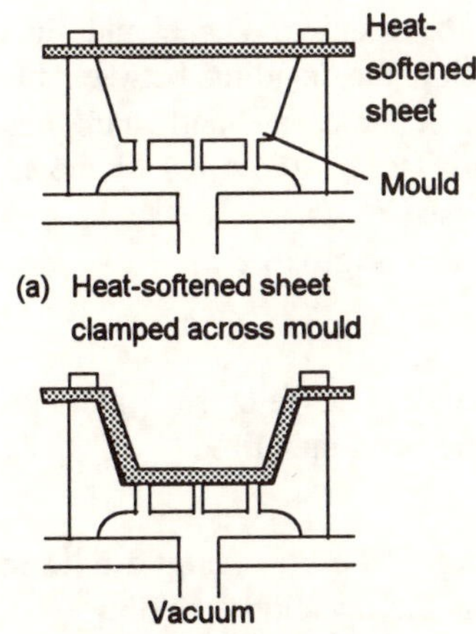

Figure 6.26 *Vacuum forming*

6.8.3 Vacuum forming

Vacuum forming is a common method of *thermoforming*. It uses a vacuum on one side of a sheet of heat-softened thermoplastic to force it against a cooled mould and hence produce the required shape (Figure 6.26). Sheets, such as 6 mm thick acrylic, are likely to be preheated in ovens before being clamped but thinner sheets of, say, 2 mm thick are likely to be heated by radiant heaters positioned over the mould. Vacuum forming can have a high output rate, but dimensional accuracy is not too good and such items as holes, threads and enclosed shapes cannot be produced. The method is used for the production objects such as open plastic containers, bath tubs and decorative panels.

6.8.4 Calendering

Calendering is a process used to form thermoplastic films, sheets and coated fabrics. The most common use has been for plasticised PVC. Calendering consists of feeding a heated paste-like mass of the plastic into the gap between two rolls, termed nip rolls. It is squeezed into a film which then passes over cooling rolls before being wound round a wind-up roll (Figure 6.27). As shown in the figure, this process can also be used to coat a fabric with a polymer.

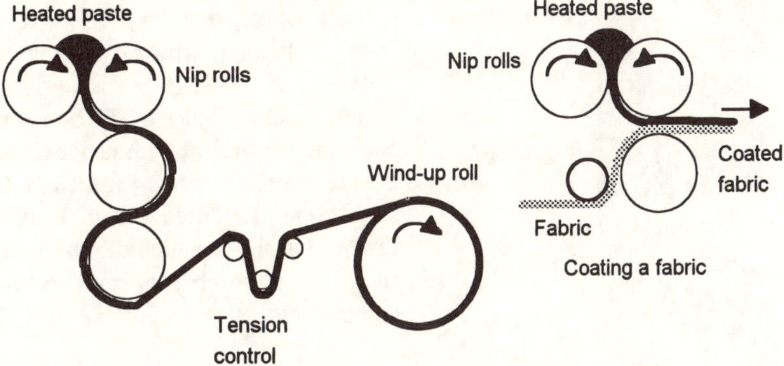

Figure 6.27 *Calendering*

6.9 Cutting

The processes used to shape a polymer generally produce the finished article with no further, or little, need for machining or any other process. With injection moulding, compression moulding and blow moulding there is a need to cut off gates and flashing; with extrusion, lengths have to be cut off. As with metals, single-point and multi-point cutting tools can be used with polymers. Where discontinuous, rather than continuous, chips are produced and the machined surface becomes excessively rough as a result of chips being sheared off. It is thus desirable to select cutting conditions which result in the formation of continuous chips.

Polymers tend to have low melting points and thus machining conditions, which do not result in high temperatures being produced, are vital if the material is not to soften and deform.

6.10 Forming processes with ceramics

In general, most ceramic products are made by moulding a powdered mass, with a binder if necessary, into the required shape and then heating it to a high temperature to develop the bonding between the particles and form the product. As most ceramics are both hard and brittle, the shaping process has generally to be the final shape as machining and further working cannot be used.

The methods used for forming ceramics can be grouped as:

1 *Wet shaping*
A wet-mixed mass is formed to the required shape by such methods as slip casting, tape casting and extrusion and then fired.

2 *Dry shaping*
This involves powders being compressed before being fired and includes die pressing, reaction bonding and injection moulding.

3 *Manufacture of glass and glass ceramics*
Glass melts are produced by heating to melting point the mixed raw materials before shaping them by some moulding machinery.

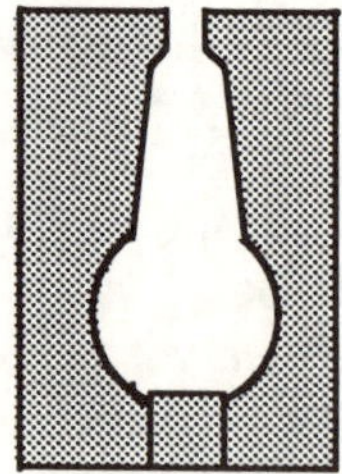

(a) Assemble mould

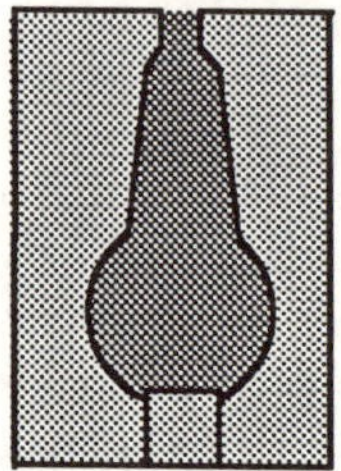

(b) Fill with slip

(c) Drain liquid

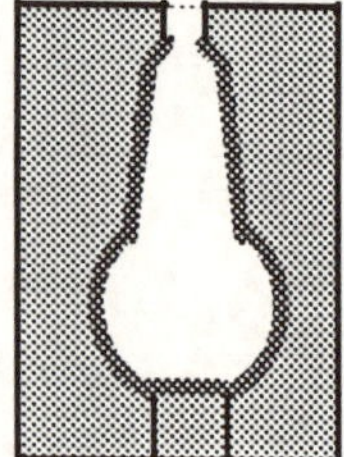

(d) Trim

(e) Remove mould and fire

Figure 6.28 *Slip casting*

6.10.1 Wet shaping

The traditional method of *wet shaping* used in the pottery trade involves the shaping of a wet clay-like mass into the required shape, e.g. on a potter's wheel, and then firing it. Such methods are used for largish shapes with axial symmetry, e.g. pots, or hand-shaped pieces.

With *slip casting*, a suspension of clay in water is poured into a porous mould (Figure 6.28). Water is absorbed by the walls of the mould and so the suspension immediately adjacent to the mould walls turns into a soft solid. When a sufficient layer has built up, the remaining suspension is poured out to leave a hollow clay form which is then removed from the mould and fired. This method is used for products such as wash basins and other sanitary ware, and thin-walled components.

6.10.2 Dry shaping

Die pressing is commonly used with oxide, carbide and nitride engineering ceramics for the formation of small shapes such as electronic ceramic components. The process is the same as that described earlier for the sintering of metals. The powdered raw material is packed into a die and then compressed. After compaction the green compact is ejected from the die and then fired to sinter the particles.

Injection moulding (the method being essentially the same as that used for polymers) is used for complex irregular shapes with holes, channels, etc.

6.10.3 Manufacture of glass and glass ceramic products

Glasses are produced by heating the appropriate raw ingredients to give a viscous melt. Sheet glass can then be made by drawing glass from the

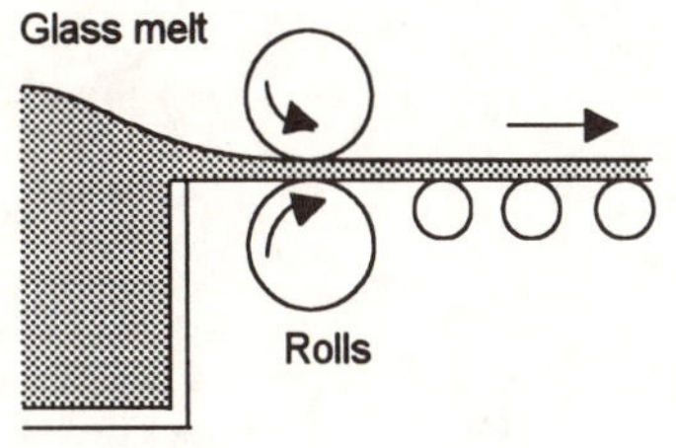

Figure 6.29 *Rolling glass*

melt and flattening it to the sheet shape by passing it between rolls (Figure 6.29). The product is not perfectly flat and parallel and does contain some imperfections. Thus most sheet glass, e.g. for windows, is now made by the *float process*; this gives flat, parallel, distortion-free glass. The float process involves the molten glass flowing in a continuous strip on the surface of molten tin (Figure 6.30).

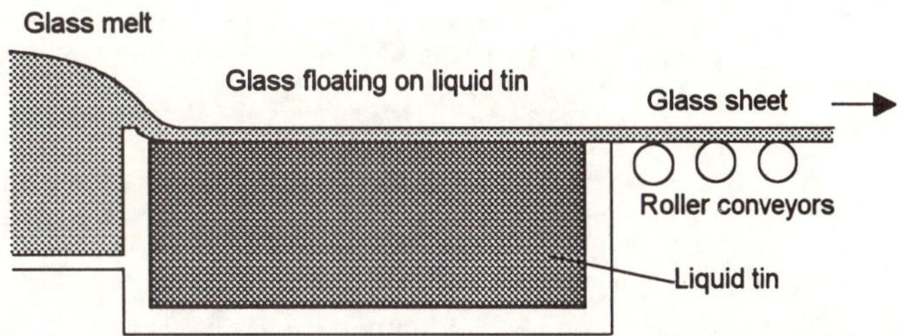

Figure 6.30 *Float glass process*

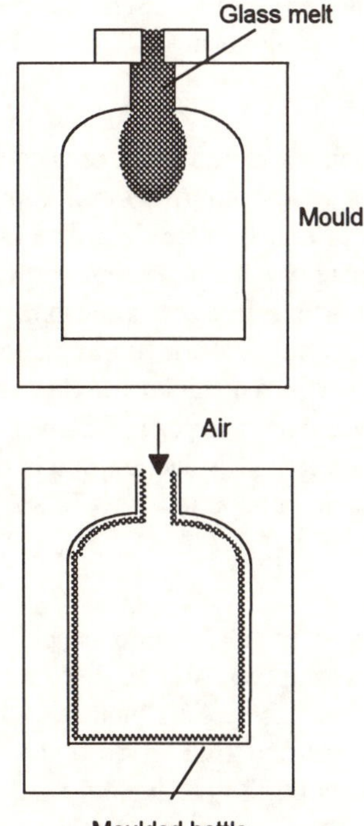

Figure 6.31 *Blow moulding*

Blow moulding can be used for the production of glass bottles and other glass containers. The glass melt, at a temperature at which its viscosity is just right, flows through an orifice and has a large drop of glass sheared off in a bottle-shaped mould and then air blown into it to cause the glass to expand out to form the bottle (Figure 6.31).

The cooling of glass from its melt temperature to room temperature is fairly rapid and results in stresses developing in the glass. To eliminate these an *annealing* process is used, the glass being heated to an appropriate temperature and then slowly cooled. For soda glass this temperature is about 500°C. At this temperature sufficient movement of the molecules is possible for the stresses to be relieved without the product losing its shape.

Where glass is to be used for its transparency it is necessary for the glass not to crystallise during its processing, the surfaces of the crystal grains scattering light and so reducing the light transmitted. *Glass ceramics*, however, are so designed that crystallisation occurs to give a fine-grained polycrystalline material. Such a material has considerably higher strength than most glasses and retains its strength to much higher temperatures. It also has excellent resistance to thermal shock. Glass ceramics are produced by using a raw material containing a large number of nuclei on which crystal growth can start, e.g. small amounts of oxides such as those of titanium, phosphorus or zirconium. The material is heated to form a glass, e.g. 1650°C, to the required product shape and then cooled. The product might be in the form of sheets which are then edge finished and surface decoration applied before being heated to a high enough temperature, e g. 900°C, to give controlled grain growth until the required grain size is obtained. Figure 6.32 illustrates this sequence for the production of cooker tops. Most forms of glass ceramic are based on $Li_2O–Al_2O_3–SiO_2$ and $MgO–Al_2O_3–SiO_2$. Such a ceramic has a very low coefficient of expansion. Glass ceramics are used for such applications as cooker tops, cooking ware and telescope mirrors.

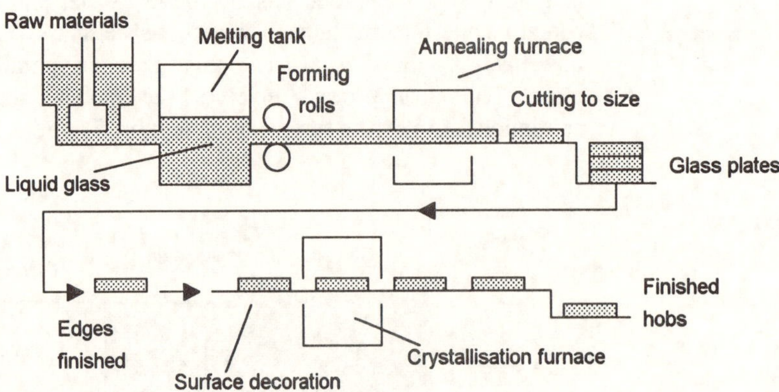

Figure 6.32 *The stages in the production of a ceramic glass cooker top*

6.11 Forming processes with composites

The following are some of the methods commonly used for the manufacture of fibre-reinforced polymer matrix composites:

1 *Hand lay-up*
The fibres might be in the form of chopped-strand mats or woven fabrics, the fibres being long or short and any required orientation. The mat or fabric is placed in a mould or on a former shaped to the form required of the finished product (Figure 6.33). A liquid thermosetting resin is then mixed with a curing agent and applied with a brush or roller to the fabric. Layers of fabric impregnated with the resin are used to build up the required thickness. Curing, i.e. waiting for the thermosetting polymer bonds to form a network, is usually at room temperature. Such a method is labour intensive but particularly suited to one-offs or small production runs of such items as the hulls of boats.

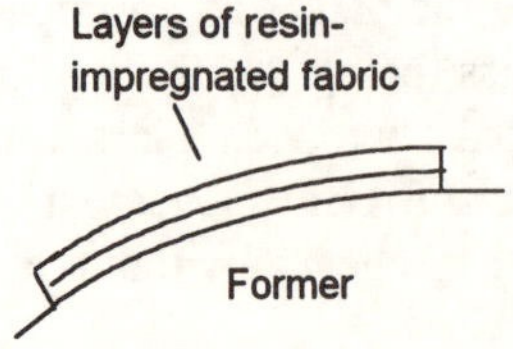

Figure 6.33 *Hand lay-up*

2 *Spray-up*
This method involves chopped fibres, resin and hardener being sprayed on to a mould. To remove trapped air, the sprayed composite has to be rolled before the resin cures. This method only involves short, randomly orientated fibres and thus gives a lower strength composite than hand lay-up, though the production rate is higher. Such a method is used for items such as sinks and baths.

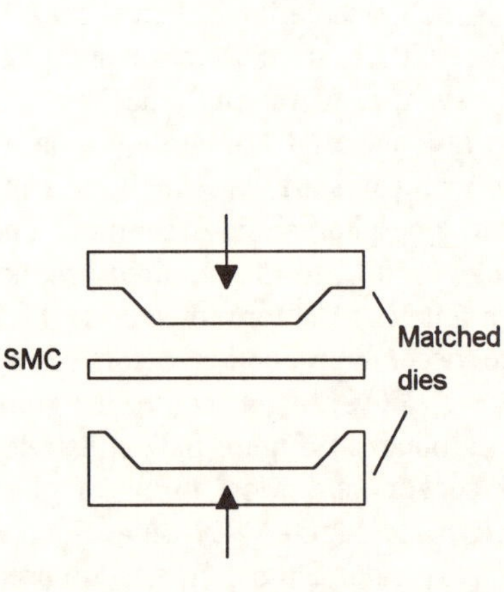

SMC

Matched dies

Figure 6.34 *Sheet moulding*

3 *Sheet moulding*
Layers of fibres are pre-impregnated with resin and partially cured, such a material being referred to as sheet moulding compound (SMC). The fibres can be long or short and orientated. These sheets have a shelf-life of some three to six months at room temperature. The sheets are stacked on the open mould surface and then forced into the mould and the required shape before being fully cured. Figure 6.34 illustrates this when a pair of matched dies are used. This forcing can also be done by vacuum forming, the air between a

sheet and a mould surface being removed and so the atmospheric pressure on the other side of the sheet forces it against the mould surface. This method is used for long production runs of such items as doors and panels.

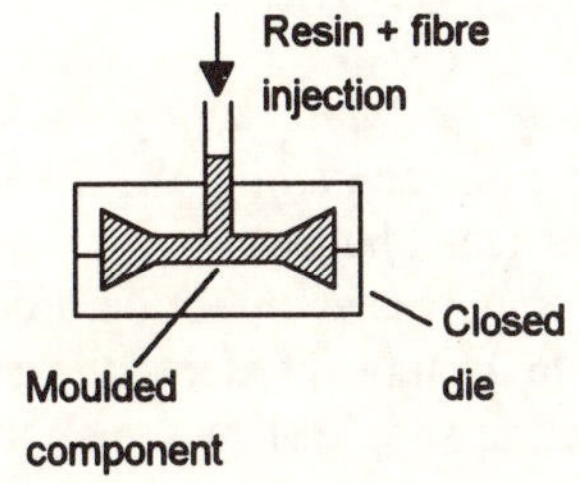

Figure 6.35 *Resin transfer moulding*

4 *Dough moulding*
This is similar to sheet moulding but involves using dough moulding compound (DMC). This is a blend of short fibres and resin that has the consistency of bread dough or putty. The DMC is pressed into a open mould and then cured.

5 *Resin transfer moulding*
This involves fibres and resin being mixed and then injected under pressure into a closed mould before being cured (Figure 6.35). This method is used for such products as fan blades, water tanks, seating, bus shelters and machine cabinets.

6 *Pultrusion*
This method is used for the production of long lengths of uniform cross-section rods, tubes or I-sections. Continuous lengths of fibre reinforcement are passed through a bath of resin and then pulled through a heated die to give the required shape product (Figure 6.36). The reinforcement may be woven fibres, a number of strands collected into a bundle with little or no twist (termed a roving), biaxial materials or random mat. Where unidirectional fibres are used, the longitudinal properties will be significantly different from the transverse properties, e.g. for 75% glass fibres in polyester, a longitudinal tensile strength of 1000 MPa and a transverse tensile strength of 40 MPa. If biaxially balanced fibres are used the strength may be of the order of 200 MPa in both longitudinal and transverse directions.

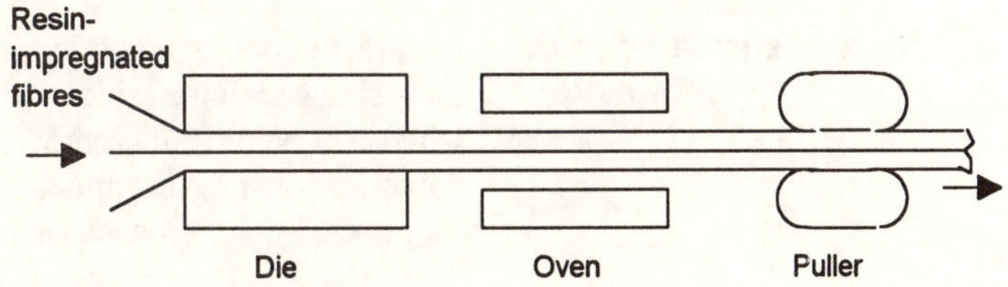

Figure 6.36 *Pultrusion*

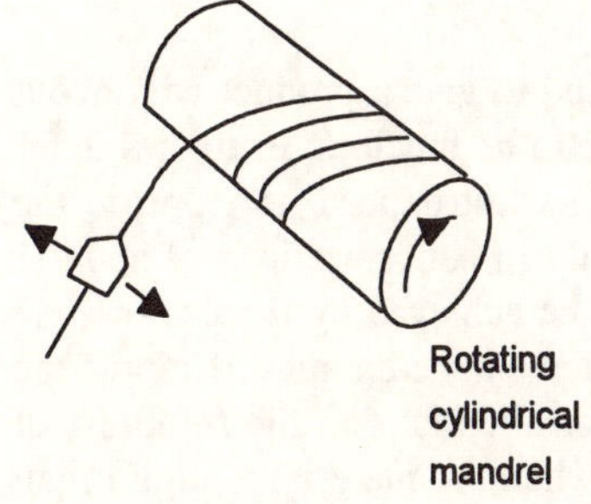

Figure 6.37 *Filament winding*

7 *Filament winding*
Continuous lengths of fibres are passed through resin and laid out in the required directions on a mandrel (Figure 6.37), this being termed *wet winding*, or pre-impregnated fibres used, this being termed *dry winding*. This can be done using a computer-controlled system so that the fibres are laid down in a predetermined manner to give the required orientations. After curing, the mandrel is removed. Products made using this method are pressure vessels, helicopter blades and storage tanks.

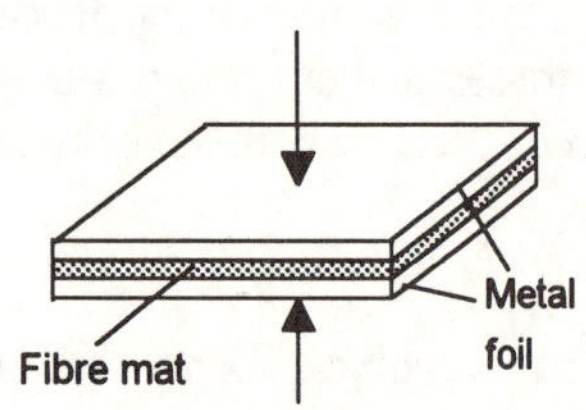

Figure 6.38 *Diffusion bonding*

6.11.1 Metal matrix composites

The methods used for manufacturing metal matrix composites can be grouped as:

1 *Solid state methods*
There are two basic methods, *diffusion bonding* and *powder processing*. With diffusion bonding, a fibre mat containing the fibres held in place by a polymer binder is sandwiched between two sheets of foil (Figure 6.38). A number of such sandwiches might be used and then the stack is hot pressed in a die to form the product. It is an expensive process and is generally limited to simple shapes such as plates and tubes. With powder processing, discontinuous fibres or particles are mixed with the metal powder and the mixture then hot pressed.

2 *Liquid state methods*
A simple form of liquid state method is just to mix the fibres or particles with the liquid metal and then cast the metal in one of the usual ways. Uniform mixing is, however, difficult to achieve. Another method is known as *squeeze casting*. This involves the die cavity having a bundle of fibres or fibre mat inserted and then molten metal poured in (Figure 6.39). Pressure is then applied and forces the metal into the fibres.

3 *Deposition methods*
With these methods, the matrix material is vapour deposited or electroplated onto the fibres, the coated fibres often then being hot pressed. An example of a vapour-deposited composite is silicon carbide reinforcement of aluminium. An example of an electroplated composite is tungsten reinforcement of nickel.

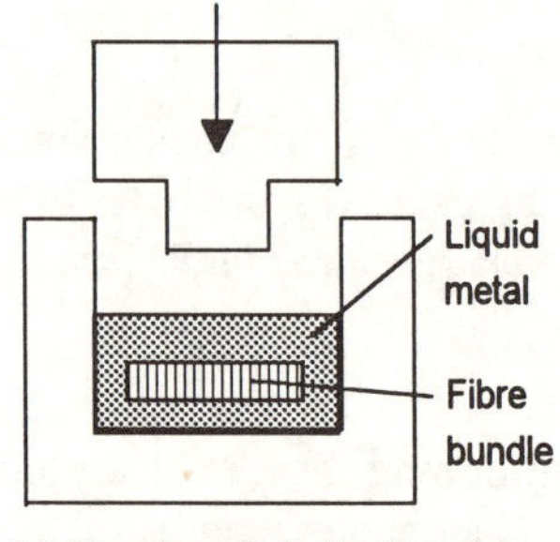

(a) Fibre bundle in liquid metal

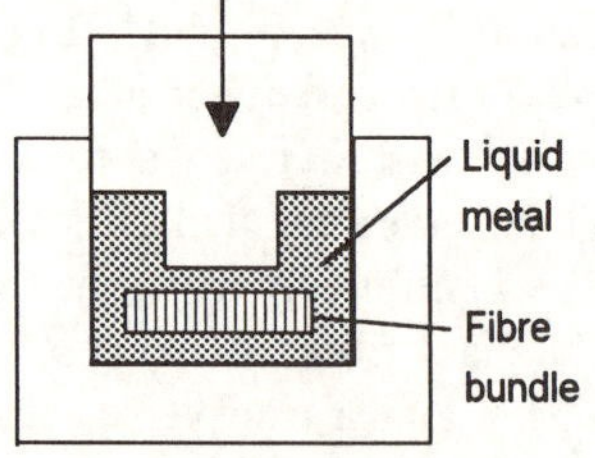

(b) Metal impregnates fibres

Figure 6.39 *Squeeze casting*

6.12 Integrated circuit fabrication

Integrated circuits involving large numbers of electronic components all on a single, minute, silicon chip are an integral feature of modern life, whether it be as a microprocessor in a computer or the control chip in a domestic washing machine. The following is an outline of the types of processes used to produce such chips.

1 *Purification of silica*
Silica can be extracted from silica sand to give a product with about 98% purity. It does, however, require to be much purer than this for use in electronics since the dopants introduced to control the properties of silicon are generally only in concentrations of about 1 part in 10^7. The required purity can be achieved by the use of *zone refining*. This involves a heating element being moved along the length of a silicon rod and is based on the fact that the solubility of impurities in the material is greater when the material is liquid than when solid. Thus by moving the zone of melting along the length of a silicon rod, so the impurities can be swept up in the molten region.

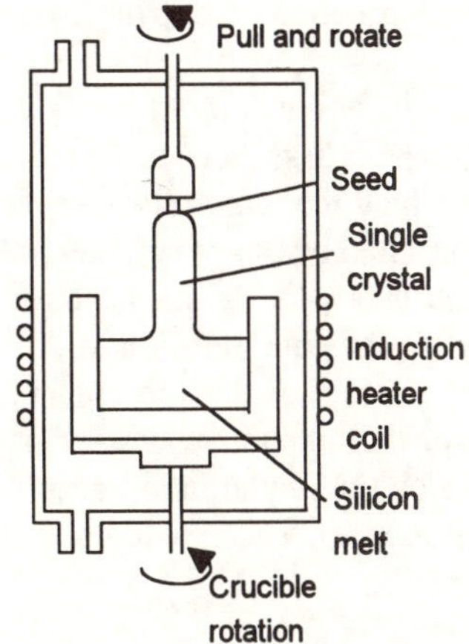

Figure 6.40 *Czochralski crystal puller*

2 *Crystal growth*
In addition to high purity, a single crystal is required, rather than a polycrystalline structure, since the dislocations at grain boundaries and the change in grain orientation can affect the electrical conductivity. A single crystal of silicon can be produced by the *Czochralski technique* (Figure 6.40). Polycrystalline silicon is heated to melting and an existing seed crystal with the required orientation is mounted above the melt surface and brought into contact with it. It is then rotated as it is slowly drawn upwards. As the melt freezes on the crystal it does so in the same orientation and a single crystal rod is gradually created.

3 *Slice preparation*
The single rod crystal is cut up into slices of about 500 to 1000 μm thickness by means of a diamond saw. The cutting action leaves some surface damage on the slices and this is removed by lapping. This involves rubbing slices between two plane parallel rotating steel discs while using a fine abrasive, usually alumina. Residual damage is then removed by etching the surfaces with a suitable chemical.

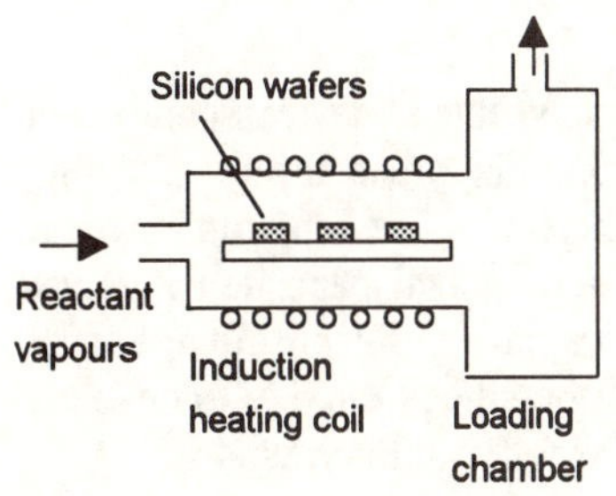

Figure 6.41 *Chemical vapour deposition*

4 *Epitaxial growth*
The surface of the slices still, however, contains defects and it is the surface layers of a silicon chip in which the circuits are 'fabricated'. *Chemical vapour deposition* (CVD) can be used to improve the surface by depositing silicon vapour onto the surface (Figure 6.41). If the temperature is above about 1100°C, the deposited layer has the same orientation as the substrate. This process is termed *epitaxy* and the deposited layer the *epitaxial layer*. The epitaxial layer can readily be doped by the inclusion of dopant atoms in the silicon vapour. In this way we can obtain a silicon chip with an n-type or p-type epitaxial layer in which we can then construct transistors and other circuit elements.

5 *Producing a silicon dioxide layer*
Silicon dioxide has a resistivity of about 10^{15} Ω m, compared with silicon with a resistivity of about 10^{3} Ω m. Thus silicon dioxide provides an electrical insulating layer which can be used as a mask for the selective diffusion of dopants into silicon. Silicon dioxide layers can be produced in a number of ways. One method is pass oxygen or water vapour (Figure 6.42) over silicon slices at about 1000°C.

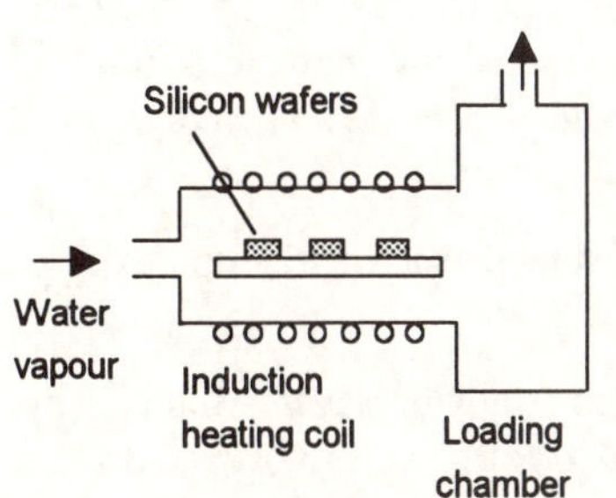

Figure 6.42 *Producing a silicon dioxide layer*

6 *Doping*
The introduction of dopant atoms into a silicon slice can be by *thermal diffusion*. For atoms to diffuse into the surface layers of a silicon slice we need a large concentration of dopant atoms at the surface of the silicon slice. This can be done by exposing the surface to a vapour containing the dopant atoms. A problem with thermal diffusion is that more than one diffusion cycle is likely to be necessary to build up an integrated circuit and the high temperature

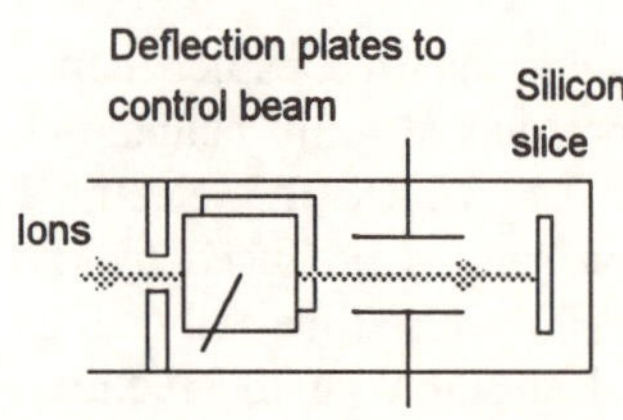

Figure 6.43 *Ion implanter*

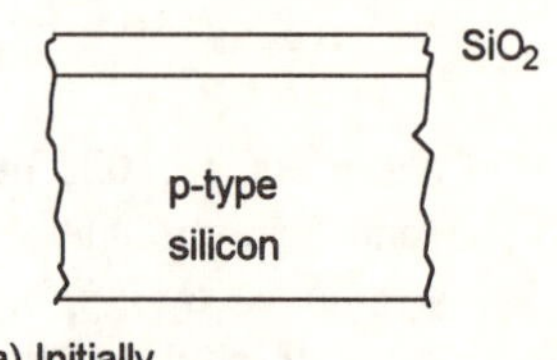

(a) Initially

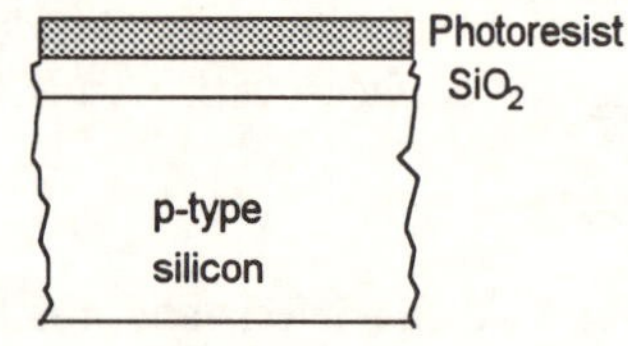

(b) Coated with photoresist

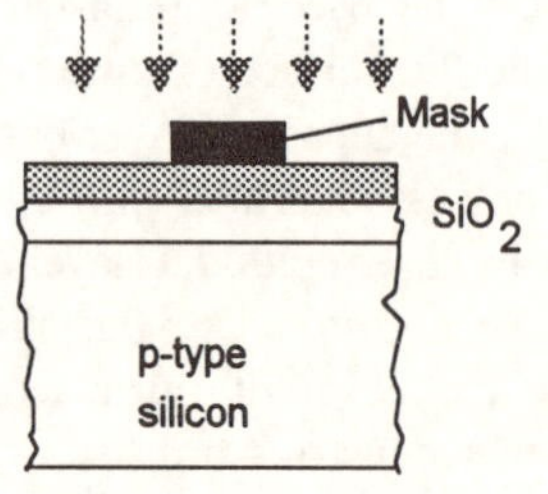

(c) Cover with mask and expose to UV

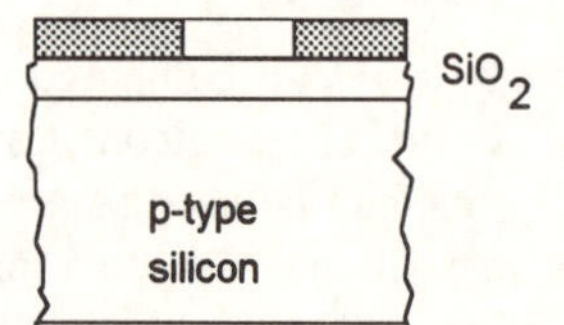

(d) Dissolve unexposed photoresist

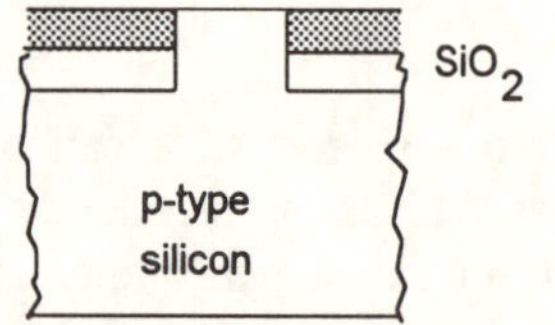

(e) Etch the uncovered silicon dioxide

Figure 6.44 *Stages in opening a window in silicon dioxide*

required for each diffusion causes a redistribution of dopant layers already laid down.

An alternative to thermal diffusion is *ion implantation.* This method involves firing the dopant atoms at high speeds into the silicon slice (Figure 6.43). The point at which the beam hits the slice and the area which it covers can be controlled by applying potential differences to the deflection plates, the depth to which the beam penetrates by controlling the velocity of the ions and the concentration of dopant implanted by controlling the ion current. This method thus allows fine control of the amount of dopant, its depth of penetration and the area in which the implantation occurs. As it is a comparatively low-temperature process, it does not excessively disturb previous processing stages.

7 *Electrical connections*
Electrical connections need to be made between components in a circuit and also between the circuit and the outside world. This can be achieved by depositing metal on the surface. Aluminium is widely used.

6.12.1 Photolithography and masking

To produce an integrated circuit we need to be able to develop patterns of n- and p-type silicon and aluminium tracks. The procedure for allowing selected parts of the surface of a silicon slice to be ion implanted or diffused with n- or p-type dopants and have metal deposited involves photolithography and masking. Thus if we have a silicon dioxide layer on silicon and wish to open a window in the silicon dioxide to enable a dopant to be introduced, the procedure is as follows (Figure 6.44):

1 The initial slice may be p-type silicon which has a surface layer of silicon dioxide (Figure 6.44(a)).

2 Coat the silicon dioxide layer with a negative photoresist. This is initially a monomer or short-chain polymer which is soluble in a solvent (Figure 6.44(b)).

3 Cover the photoresist with a mask which contains the pattern to be transferred to the photoresist. Then expose the photoresist to ultraviolet light (Figure 6.44(c)). Exposure of the negative photoresist to ultraviolet light results in polymerisation or cross-linking of the short chains and a considerable reduction in solubility in the solvent.

4 Develop the pattern in the photoresist by dissolving the non-exposed resist in a solvent (Figure 6.44(d)).

5 Apply an etch to remove the silicon dioxide layer which is not covered by the photoresist (Figure 6.44(e)). Hydrofluoric acid is the standard etch used for silicon dioxide.

6 The result is a window in the silicon dioxide through which we might introduce, for example, n-type dopant.

7 The above procedure can be repeated many times in order to build up the required circuit and components. As an illustration, Figure 6.45 shows the layers that can be used for a resistor. The aluminium provides the connections to the ends of the resistor which have been heavily doped and has resulted in such a large number of negative charge carriers that it acts as a good conductor and also provides a region of good contact between the aluminium and the n-type material. The resistance element is the n-type strip. This might be a channel about 100 μm long and 10 μm wide.

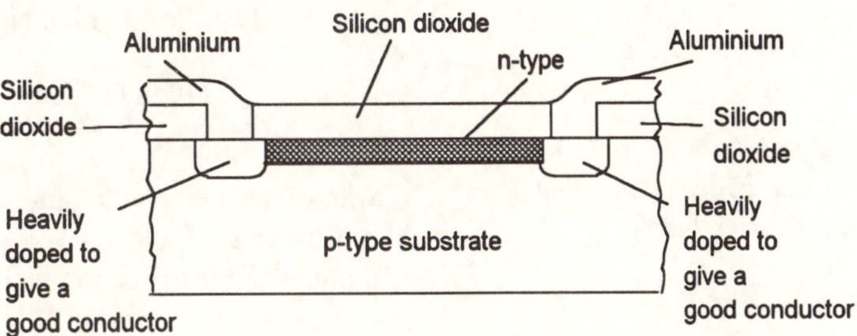

Figure 6.45 *An integrated resistor*

6.13 Selection of process

In selecting the process to be used for a product, the following have to be considered:

1 *What is the material?*
The type of material to be used affects the choice of processing method, e.g. a ductile material.

2 *What is the shape?*
The shape of the product is generally a vital factor in determining which type of process can be used, e.g. a tube.

3 *What type of detail is involved?*
Is the product to have holes, threads, inserts, hollow sections, fine detail, etc.?

4 *What dimensional accuracy and tolerances are required?*
High accuracy would rule out some methods such as sand casting.

5 *Are any finishing processes to be used?*
Is the process used to give the product its final finished state or will there have to be an extra finishing process?

6 *What quantities are involved?*
Is the product a one-off, a small batch, a large batch or continuous production? While some processes are economic for small quantities, others do not become economic until large quantities are involved. Thus, for example, open die forging could be economic for small numbers but closed die forging would not be economic unless large numbers were produced.

6.13.1 Surface finish

Roughness is defined as the irregularities in the surface texture and takes the form of a series of peaks and valleys which may vary in both height and spacing and is a characteristic of the process used. One measure of roughness is the *arithmetical mean deviation*, denoted by the symbol R_a. This is the arithmetical average of the variation of the profile above and below a reference line throughout the prescribed sampling length. The reference line may be the centre line, this being a line chosen so that the sums of the areas contained between it and those parts of the surface profile which lie on either side of it are equal (Figure 6.46). Thus:

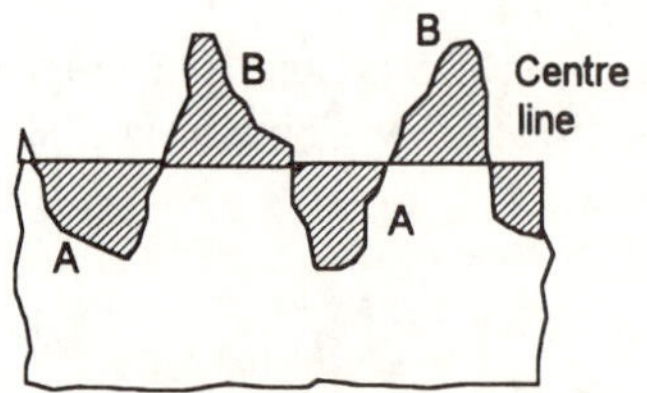

Figure 6.46 *Area between surface above centre line equals area below*

$$R_a = \frac{\text{sum of areas A} + \text{sum of areas B}}{\text{sample length}} \times 1000$$

where the sample length is in millimetres and the areas in square millimetres. Table 6.4 indicates the relation of R_a values to surface texture; Table 6.5 shows typical values for different processes.

Table 6.4 *Relation of R_a values to surface texture*

Surface texture	Roughness R_a μm
Very rough	50
Rough	25
Semi-rough	12.5
Medium	6.3
Semi-fine	3.2
Fine	1.6
Coarse-ground	0.8
Medium-ground	0.4
Fine-ground	0.2
Super-fine	0.1

Table 6.5 *Roughness values for different processes*

Process	Roughness R_a μm
Sand casting	25 to 12.5
Hot rolling	25 to 12.5
Sawing	25 to 3.2
Planing, shaping	25 to 0.8
Forging	12.5 to 3.2
Milling	6.3 to 0.8
Boring, turning	6.3 to 0.4
Extruding	3.2 to 0.8
Cold rolling	3.2 to 0.8
Drawing	3.2 to 0.8
Die casting	1.6 to 0.8
Grinding	1.6 to 0.1

The degree of roughness that can be tolerated for a component depends on its use. Thus, for example, precision sliding surfaces will require R_a values of the order of 0.2 to 0.8 μm with more general sliding surfaces 0.8 to 3 μm. Gear teeth are likely to require R_a values of 0.4 to 1.6 μm, friction surfaces such as clutch plates 0.4 to 1.5 μm, mating surfaces 1.5 to 3 μm.

6.14 Characteristics of metal forming processes

The following is a discussion of the characteristics of the various processes used for metal forming and the types of products that can be obtained from them.

6.14.1 Casting of metals

Casting can be used for components from masses of about 10^{-3} kg to 10^4 kg, with wall thicknesses from about 0.5 mm to 1 m. Castings need to have rounded corners, no abrupt changes in section and gradual sloping surfaces. Casting is likely to be the optimum method in the circumstances listed below but not for components that are simple enough to be extruded or deep drawn.

1. *The part has a large internal cavity*
   There would be a considerable amount of metal to be removed if machining was used; casting removes this need.
2. *The part has a complex internal cavity*
   Machining might be impossible; by casing, however, very complex internal cavities can be produced.
3. *The part is made of a material which is difficult to machine*
   The hardness of a material may make machining very difficult, e.g. white cast iron, but this presents no problem with casting.
4. *The metal used is expensive and so there is to be little waste*
   Machining is likely to produce more waste than occurs with casting.
5. *The directional properties of a material are to be minimised*
   Metals subject to a manipulative process often have properties which differ in different directions.
6. *The component has a complex shape*
   Casting may be more economical than assembling a number of individual parts.

An important consideration in deciding whether to use casting and which casting process to use is the tooling cost for making the moulds. When many identical castings are required, a method employing a mould which may be used many times will enable the mould cost to be spread over many items and may make the process economic. Where just a one-off product is required, the mould used must be as cheap as possible since the entire cost will be defrayed against the single product.

Table 6.6 *Casting processes*

Process	Usual materials	Section thickness mm	Size kg	Roughness R_a μm	Production rate, items per hour
Sand casting	Most	> 4	0.1–200 000	25 to 12.5	1–60
Gravity die casting	Non-ferrous	3 to 50	0.1–200	3.2 to 1.6	5–100
Pressure die casting: high pressure	Non-ferrous	1 to 8	0.0001–5	1.6 to 0.8	Up to 200
Pressure die casting: low pressure	Non-ferrous	2 to 10	0.1–200	1.6 to 0.8	Up to 200

Each casting method has important characteristics which determine its appropriateness in a particular situation. Table 6.6 illustrates some of the key differences between the casting methods. The following factors largely determine the type of casting process used:

1 *Large heavy casting*
Sand casting can be used for very large castings.

2 *Complex design*
Sand casting is the most flexible method and can be used for very complex castings.

3 *Thin walls*
Pressure die casting can cope with walls as thin as 1 mm. Sand casting cannot cope with such thin walls.

4 *Good reproduction of detail*
Pressure die casting gives good reproduction of detail, sand casting being the worst.

5 *Good surface finish*
Pressure die casting gives the best finish, sand casting being the worst.

6 *High melting point alloys*
Sand casting can be used.

7 *Tooling cost*
This is highest with pressure die casting. Sand casting is cheapest. However, with large number production, the tooling costs for metal moulds can be defrayed over a large number of castings, whereas the cost of the mould for sand casting is the same no matter how many castings are made since a new mould is required for each casting.

6.14.2 Manipulation of metals

Manipulative methods involve the shaping of a material by means of plastic deformation methods. Depending on the method, components can be produced from as small as about 10^{-5} kg to 100 kg, with wall thicknesses from about 0.1 mm to 1 m. Table 6.7 shows some of the characteristics of the different processes.

Table 6.7 *Manipulative processes*

Process	Usual materials	Section thickness mm	Minimum size	Maximum size	Roughness R_a μm	Production rate, item per hour
Closed-die forging	Steels, Al, Cu, Mg alloys	3 upwards	10 cm^2	7000 cm^2	3.2 to 12.5	Up to 300
Roll forming	Any ductile material	0.2 to 6			0.8 to 3.2	
Drawing	Any ductile material	0.1 to 25	3 mm dia.	6 m dia.	0.8 to 3.2	Up to 3000
Impact extrusion	Any ductile material	0.1 to 20	6 mm dia.	0.15 m dia.	0.8 to 3.2	Up to 2000
Hot extrusion	Most ductile materials	1 to 100	8 mm dia.	500 mm dia.	0.8 to 3.2	Up to 720 m
Cold extrusion	Most ductile materials	0.1 to 100	8 mm dia	4 m long	0.8 to 3.2	Up to 720 m

Note: ductile materials are commonly aluminium copper and magnesium alloys and to a lesser extent carbon steels and titanium alloys.

Compared with casting, wrought products tend to have a greater degree of uniformity and reliability of mechanical properties. The manipulative processes do, however, tend to give a directionality of properties which is not the case with casting. Manipulative processes are likely to be the optimum method for product production when:

1 *The part can be be formed from sheet metal*
Depending on the form required, shearing, bending or drawing may be appropriate if the components are not too large.

2 *Long lengths of constant cross-section are required*
Extrusion or rolling would be the optimum methods in that long lengths of quite complex cross-section can be produced without any need for machining.

3 *The part has no internal cavities*
Forging can be used when there are no internal cavities, particularly if better toughness and impact strength is required than are obtainable with casting. Also, directional properties can be imparted to the material to improve its performance in service.

4 *Seamless cup-shaped objects or cans are required*
Deep drawing can be used.

6.14.3 Powder processes

Powder processes enable large numbers of small components to be made at high rates of production, for small items up to 1800 per hour, and with little, if any, finishing machining required. It enables components to be made with all materials and in particular with those which otherwise cannot easily be otherwise easily processed, e.g. the high melting point metals of molybdenum, tantalum and tungsten, and where there is a need for some specific degree of porosity, e.g. porous bearings to be oil filled.

The mechanical compaction of powders only, however, permits two-dimensional shapes to be produced, unlike casting and forging. In addition the shapes are restricted to those that are capable of being ejected from the die. Thus, for example, reverse tapers, undercuts and holes at right angles to the pressing direction have to be avoided. On a weight-for-weight basis, powdered metals are more expensive than metals for use in manipulative or casting processes; however, this higher cost may be offset by the absence of scrap, the elimination of finishing machining and the high rates of production. The maximum size for products is about 4.5 kg.

6.15 Characteristics of polymer forming processes

Injection moulding and extrusion are the most widely used processes. Injection moulding is generally used for the mass production of small items, often with intricate shapes. Extrusion is used for products which are required in continuous lengths or which are fabricated from materials of constant cross-section. The following are some of the factors involved in choosing a process:

1 *Rate of production*
Cycle times are typically: injection moulding and blow 10–60 s, compression moulding 20–600 s, thermoforming 10–60 s).

2 *Costs*
Injection moulding requires the highest capital investment, with extrusion and blow moulding requiring less capital. Compression moulding, transfer moulding and thermoforming require the least capital investment. Table 6.8 indicates the minimum output that is likely to be required to make processes economic.

Table 6.8 *Economic output*

Process	Economic output number
Machining	From 1 to 100 items
Sheet forming	From 100 to 1000 items
Extrusion	Length 300 to 3000 m
Blow moulding	From 1000 to 10 000 items
Injection moulding	From 10 000 to 100 000 items

3 *Surface finish*
Injection moulding, blow moulding, thermo-forming, transfer and compression moulding all give very good surface finishes. Extrusion gives only a fairly good surface finish.

4 *Dimensional accuracy*
Injection moulding and transfer moulding are very good, Compression moulding good and extrusion is fairly poor.

5 *Item size*
Injection moulding and machining are the best for very small items. Section thicknesses of the order of 1 mm can be obtained with injection moulding, forming and extrusion.

6 *Enclosed hollow shapes*
Blow moulding and rotational moulding can be used.

7 *Intricate, complex shapes*
Injection moulding, blow moulding, transfer moulding and casting can be used.

8 *Threads*
Threads can be produced with injection moulding, blow moulding, casting and machining.

Table 6.9 shows the processing methods that are used for some commonly used thermoplastics and Table 6.10 for thermosets.

Table 6.9 *Processing methods for thermoplastics*

Polymer	Extrusion	Injection moulding	Extrusion blow moulding	Thermo-forming	Bending and joining	As film
ABS	*	*		*	*	
Acrylic	*	*		*	*	
Cellulosics	*	*		*		*
Polyacetal	*	*	*			
Polyamide	*	*				*
Polycarbonate	*	*	*	*	*	
Polyester	*	*				
Polyethylene HD	*	*	*		*	*
Polyethylene LD	*	*	*		*	*
Polyethylene terephthalate	*	*	*			*
Polypropylene	*	*	*	*	*	*
Polystyrene	*	*	*	*	*	*
Polysulphone	*	*		*		
PVC	*	*	*	*	*	*

Table 6.10 *Processing methods for thermosets*

Polymer	Compression moulding	Transfer moulding	Laminate	Foam	Film
Epoxy			*	*	
Melamine formaldehyde	*	*	*		
Phenol formaldehyde	*	*	*	*	
Polyester	*	*	*		*
Urea formaldehyde	*	*	*		

Activity

Polythene bags are generally made by extruding the polyethylene through a die to give an extruded cylinder. The cylinder, while hot, is then inflated by air pressure. What type of structure might be expected within the bag material? Cut strips in directions along the length of the bag and at right angles and pull them between your hands. On the basis of your observations suggest the form of the structure of the material.

Problems

1. Explain why copper wires that are used for electrical conductors are usually finished by a cold drawing process and then heated to about 700°C?
2. What microstructure, and hence what properties, would you expect in cold drawn wire if there is no further treatment of it?
3. Which type of casting, sand or die casting, will produce a product with the smallest grains?
4. With a thermoplastic polymer, how does the rate of cooling from the liquid state affect the degree of crystallinity?
5. What would you expect to be the internal structure of an extruded thermoplastic?
6. The thin plastic container used to hold biscuits or chocolates in boxes of biscuits or chocolates is produced by thermoforming a thermoplastic. What is likely to be the resulting molecular structure in the various parts of the container?
7. Polypropylene twine consists of fibres of polypropylene which have been cold drawn. What is the effect of this process?
8. An aluminium-manganese alloy is found to have the following properties. Explain how they arise in terms of the structure of the alloy.

	Strength MPa	Hardness HB	Elongation %
Annealed	110	28	30
Fully work hardened	200	55	4

9. Describe how the mechanical properties of a cold-worked metal changes as its temperature is raised from room temperature to about $0.6T_m$, where T_m is the melting point on the kelvin scale.
10. How does the temperature at which working is carried out determine the grain size and so the mechanical properties?
11. Why are the mechanical properties of a cold rolled metal different in the direction of rolling from those at right angles to this direction?
12. How does a cold-rolled product differ from a hot-rolled one?
13. Brasses have recrystallisation temperatures of the order of 400°C. Roughly, what temperature should be used for the hot extrusion of brass?

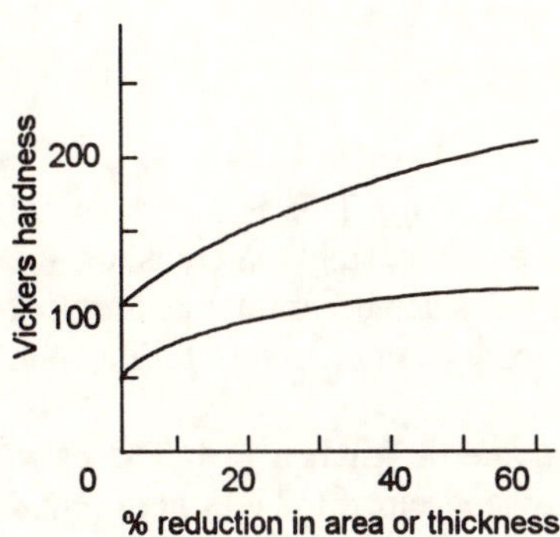

Figure 6.47 *Problem 15*

14 A brass, 65% copper and 35% zinc, has a recrystallisation temperature of 300°C after being cold worked so that the cross-sectional area has been reduced by 40%.

(a) How will further cold working change the structure and properties of the brass?

(b) To what temperature should the brass be heated to give stress relief?

(c) To what temperature should the brass be heated to anneal it?

15 Use Figure 6.46 for this problem. According to this figure:

(a) What is the maximum hardness possible with cold-rolled copper?

(b) Copper plate, already cold worked 10%, is further cold worked 20%. By approximately how much will the hardness change?

(c) Mild steel is to be rolled to give thin sheets. This involves a 70% reduction on sheet thickness. What treatment would be suitable to give this reduction and a final product which was no harder than 150 HV?

16 Suggest a method that could be used to produce a small, simple, silicon nitride ceramic former for an electronic component.

17 Suggest methods that could be used to (a) make a one-off boat hull with glass fibre-reinforced polymer, (b) mass produce door panels for lorries in glass fibre-reinforced polymer.

18 Suggest methods that could be used to make (a) thermoplastic door handles, (b) thermoplastic bath tubs, (c) thermoset dishes.

19 Suggest methods that could be used to economically manufacture (a) large numbers of piston rods, (b) one-off prototype pump housing, (c) large numbers of cans.

7 Selection

7.1 The requirements

What functions does a product have to perform? This is an important question that requires an answer before either the materials or the forming processes for the product are considered. From this stems a sequence of further questions. The following example serves to illustrate how the sequence might develop.

Consider the problem of making a domestic kitchen pan. The basic functions of a domestic kitchen pan may be deemed to be to hold liquid and allow it to be heated to temperatures of the order of 100°C. From a consideration of the function we can arrive at the basic design requirements. Thus, a consequence of these functions for the pan are the requirements for a particular shape of container which must not deform when heated to these temperatures. It must be a good conductor of heat. It must be leak proof. It must not ignite when in contact with a flame or hot electrical element. In addition there may be other requirement which are not so essential, but certainly desirable. For the pan we might thus require an attractive surface finish.

From these requirements we can now define the required properties of the materials. Thus the requirement that the material be a good conductor of heat would seem to reduce the consideration to metals, particularly when taken together with the requirement that the material can be put in contact with a flame and contain hot liquids. This would effectively rule out polymers. But what properties are required of the metal? The shape of the pan would suggest that a deep drawing process be used. As this is a cold working process then there will be a good surface finish. For deep drawing the initial material must be reasonably ductile and available in sheet form. Thus we might consider an aluminium alloy. If we look up tables, a possibility would seem to be an aluminium-magnesium alloy, AA5005. Another possibility would be a stainless steel, stainless because rusty pans would not be very desirable. If we look up tables, a possibility would seem to be 302S31. The deciding factor is likely to be cost, though there may be some prestige value attached to having stainless steel plans as opposed to aluminium which would permit a higher price to be charged. For the same volume of material, the stainless steel will probably cost about three times the aluminium alloy.

The above represents one line of argument regarding the design of pans. It is instructive to examine a range of pans and consider the materials used and what reasons might be advanced for them being chosen. Why for example are some pans made of glass, some of a ceramic, some of a steel coated on the outside with an enamel and on the inside with a non-stick polymer polytetrafluoroethylene (PTFE)?

The above is only the consideration of the container part of the pan. There is still the handle to consider. The function required is that it can be used to lift the pan and contents, even when they are hot. The properties required are thus poor thermal conductivity, able to withstand the temperatures of the hot pan, stiffness and adequate strength. The handle can be considered to be essentially a cantilever with a load, the pan and contents, at its free end. Before going so far with considering the design and materials for the handle, British Standards should be consulted. BS 6743 gives a standard specification for the performance of handles and handle assemblies attached to cookware. This sets the levels of safe performance against identified tests simulating hazards experienced in normal service. Taking this into consideration, then the need for the handle to have low thermal conductivity indicates that metal would not be a good choice. The possibility is thus a polymer. It needs however to be able to withstand a temperature of the order of 100°C at the pan end, have a reasonably high modulus of elasticity, and reasonable strength. These requirements suggest a thermoset is more likely to be feasible than a thermoplastic. A possibility is phenol formaldehyde (Bakelite). The dark colour of this material is no problem in these circumstances. When filled with wood flour, it has a high enough maximum service temperature of about 150°C, a tensile modulus of 5.0 to 8.0 GPa (high for a polymer), and a tensile strength of 40 to 55 MPa. Because it is a thermoset then the processing method could be compression moulding.

In the above considerations of the pan and the handle the item that has so far not been discussed is the life of the items. The purchaser of the pan wants it to last, without problems, for a reasonable period of time. This is likely to be years. The handle should not break during this time or discolour or deteriorate when used and washed a large number of times. The pan should not wear thin or change its mechanical properties with frequent heating, exposure to hot liquids and washing up liquids.

7.1.1 Stages in the selection process

As the above examples indicate, there are a number of stages involved in arriving at possible materials and processing requirements for a product. These can be summarised, in very simple terms, as follows:

1 Define the functions required of the product

2 Consider a tentative design, taking into account any Codes of Practice, National or International Standards

3 Define the properties required of the materials

4 Identify possible materials, taking into account availability in the required forms

5 Identify possible processes which would enable the design to be realised

6 Consider the possible materials and possible processes and arrive at a proposal for both. If not feasible, consider again the design and go back through the cycle

7 Consider how the product will behave during its service life

Activity

Design a testing procedure that could be used by the manufacturer of a new design of domestic kitchen pan to determine its behaviour during service.

7.1.2 Costs

The total cost to the consumer of a manufactured article in service, i.e. the so-called *total life cost*, is made up of a number of items. These are:

1 *The purchase price*
This includes the costs of production, the fixed costs arising from factory overheads, administration, etc. and the manufacturer's profit. The costs of production include the cost of the basic materials and the cost of manufacture

2 *The cost of ownership*
This includes maintenance, repair and replacement costs

7.2 Failure

Failures in service can arise from:

1 Errors in the original design, e.g. the wrong material used

2 The material used is in some way defective, e.g. the specification of the material to be obtained from the suppliers was not tight enough or was below specification and not detected by inspection

3 Defects are introduced during the manufacturing process, e.g. heat treatment gives cracks as a result of quenching or perhaps incorrect assembly leading to misalignment and high stresses

4 Deterioration in service, e.g. the product is exposed to unexpected corrosive environments or perhaps a temperature which results in changes in the microstructure of the material or perhaps poor maintenance which leads to nicks and gouges which act as stress raisers and a greater chance of failure due to fatigue

7.2.1 The causes of failure

Failure can arise from a number of causes, e.g.

1 The stress level is just too high and the material yields and then breaks (Figure 7.1(a)). With a ductile fracture there will have been quite significant yielding before the fracture occurs, with brittle fracture virtually none.

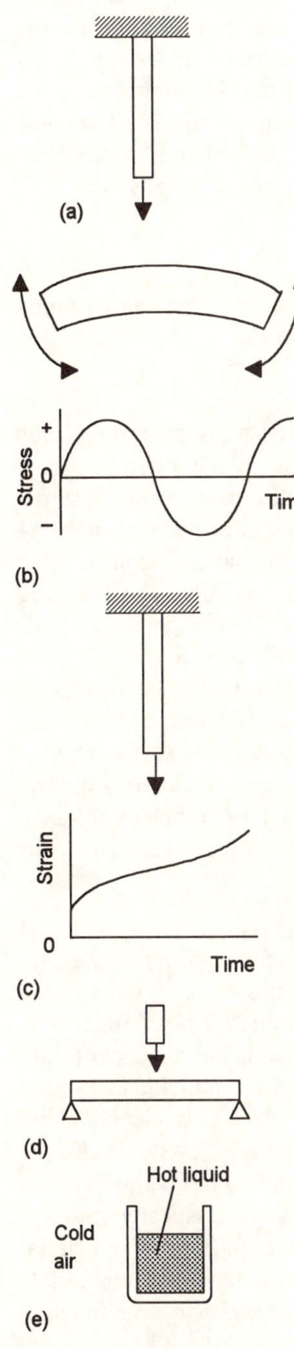

Figure 7.1 *Some modes of failure: (a) static stress, (b) alternating stresses, (c) creep, (d) impact loading, (e) differential expansion*

2 The material is subject to an alternating stress which results in fatigue failure. If you take a stiff piece of metal or plastic, and want to break it, then you will more likely flex the strip back and forth repeatedly, i.e. subject it to an alternating stress going from tension to compression to tension to compression and so on (Figure 7.1(b)). This is generally an easier way of causing the material to fail than applying a direct pull. The chance of fatigue failure occurring is increased the greater the amplitude of the alternating stresses. Stress concentrations produced by holes, surface defects and scratches, sharp corners, sudden changes in section, etc. can all help to raise the amplitude of the stresses at a particular point in the material and reduce its fatigue resistance.

3 The material is subject to a load which initially does not cause failure but the material gradually extends over a period of time until it fails. This is known as creep (Figure 7.1(c)). For most metals creep is negligible at room temperature but can become pronounced at high temperatures. For plastics, creep is often quite significant at ordinary temperatures and even more pronounced at higher temperatures. The creep of a metal, or a polymer, is determined by its composition and the temperature. For example, aluminium alloys will creep and fail at quite low stresses when the temperature rises above 200°C, while titanium alloys can be used at much higher temperatures before such a failure occurs.

4 A suddenly applied load (Figure 7.1(d)) causes failure, i.e. the energy at impact is greater than the impact strength of the material. Brittle materials have lower impact strengths than ductile materials.

5 The temperature changes and causes the properties to change in such a way that failure results, e.g. the temperature of a steel may drop to a level at which the material turns from being ductile to brittle and then easily fails as a result of perhaps an impact load.

6 A temperature gradient is produced and causes part of the product to expand more than another part, the resulting stresses resulting in failure, e.g. if you pour hot water into a cold glass then the inside of the glass tries to expand while the outside does not (glass has a low thermal conductivity) and the glass can crack (Figure 7.1(e)).

7 The product is made of materials with differing coefficients of expansion. Thus when the temperature rises the different parts expand by different amounts, with the result that internal stresses are set up. These can result in distortion and possible failure.

8 Thermal cycling in which the temperature of the product repeatedly fluctuates will result in cycles of thermal expansion and contraction. If the material is constrained in some way then internal stresses will be set up and as a consequence the thermal cycling will result in alternating stresses being applied to the material. Fatigue failure can result. This is sometimes referred to as thermal fatigue.

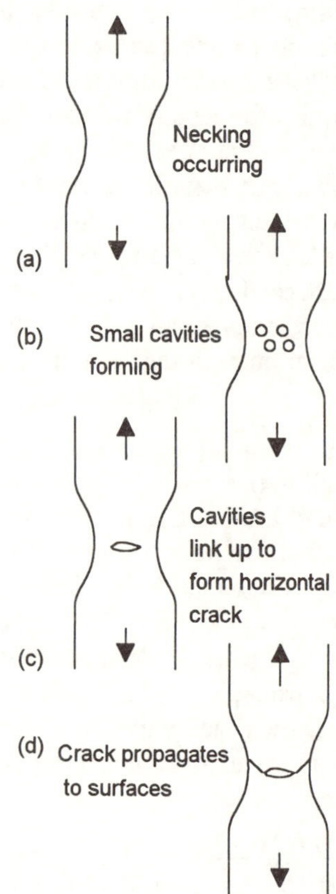

Figure 7.2 *Stages in ductile fracture*

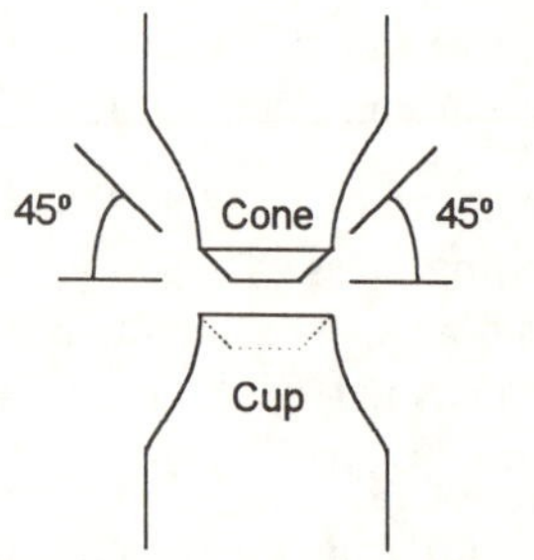

Figure 7.3 *Ductile failure*

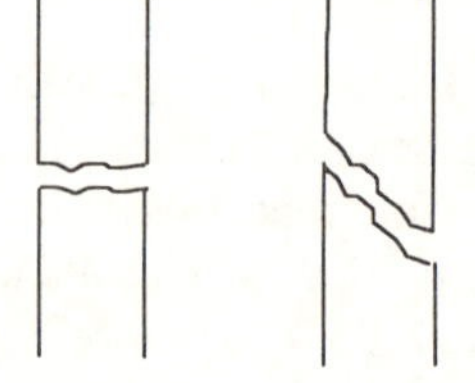

Figure 7.4 *Brittle fracture*

9 Degradation because of the environment in which the material is situated, e.g. corrosion of iron leading to a reduction in the cross-section of a product and hence the resulting increase in stress leading to failure. Corrosion prevention, e.g. painting iron, is a major way of avoiding this type of problem. Plastics may become brittle as a result of exposure to the ultraviolet radiation in sun light. This effect can be reduced by adding stabilisers to the polymer.

7.2.2 Examination of failures

The following are illustrations of the types of failure that can be found with metals, polymers, ceramics and composites:

1 *Ductile failure with metals*
A ductile material is characterised by having a significant plastic region to its stress-strain graph. When a ductile material has a gradually increasing tensile stress applied then, when yielding starts, the cross-sectional area of the material becomes reduced, necking being said to occur (Figure 7.2). Eventually after a considerable reduction in cross-sectional area the material fails. The resulting fracture surfaces show a cone and cup formation (Figure 7.3). This occurs because, under the action of the increasing stress, small internal cracks form which gradually grow in size until there is an internal, almost horizontal, crack. The final fracture occurs when the material shears at an angle of 45° to the axis of the applied stress. Such a type of failure is referred to as a *ductile fracture*. Materials can also fail in a ductile manner in compression, such fractures resulting in a characteristic bulge and series of axial cracks around the edge of the material.

2 *Brittle failure with metals*
A brittle material has virtually no plastic region to its stress-strain graph. Thus when a brittle material fractures there is virtually no plastic deformation. Figure 7.4 shows possible forms of fracture in such a situation. The surfaces of the fractured material appear bright and granular due to the reflection of light from individual crystals; the fracture has grains within the material which have cleaved along planes of atoms. We can consider the sequence of events leading to brittle fracture to be that when stress is applied the bonds between atoms and between grains in the material are elastically strained, then at some critical stress the bonds break, remember the material is brittle and there is no plastic deformation and hence small-scale slip, and a crack propagates through the material to give fracture. Such failure is known as *brittle fracture*.

3 *Fatigue failure with metals*
Fatigue failure often starts at some point of stress concentration. This point of origin of the failure can be seen on the failed material as a smooth, flat, semicircular or elliptical region, often referred to as the nucleus. Surrounding the nucleus is a burnished zone with ribbed markings. This smooth zone is produced by the crack

propagating relatively slowly through the material and the resulting fractured surfaces rubbing together during the alternating stressing of the component. When the component has become so weakened by the steadily spreading crack that it can no longer carry the load, the final abrupt fracture occurs. This region of abrupt failure has a crystalline appearance. Figure 7.5 shows the various stages in the growth of a fatigue failure.

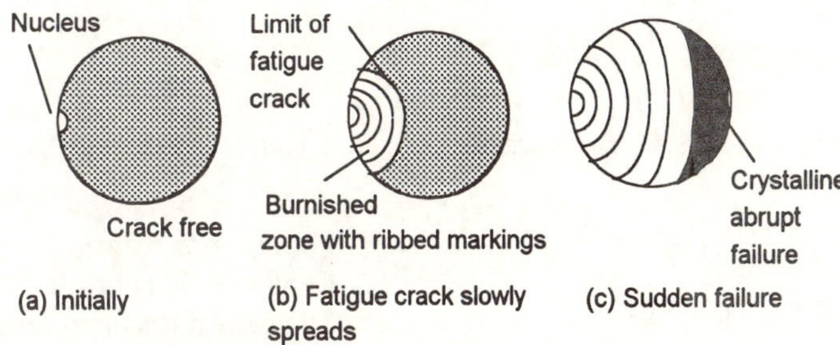

Figure 7.5 *Stages in fatigue failure*

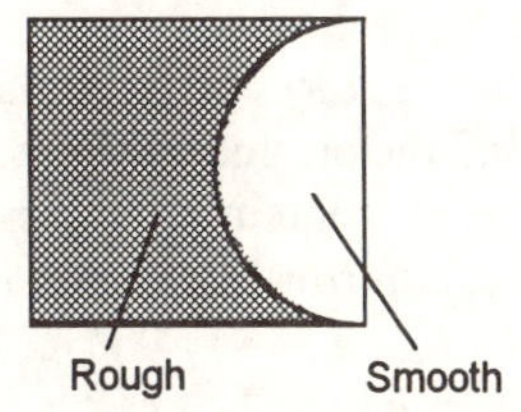

Figure 7.6 *Fracture surface for a brittle polymer*

4 *Failure with polymers*
Brittle failure with polymeric materials is a common form of failure with materials below their glass transition temperature, i.e. amorphous polymers. The resulting fracture surfaces show a mirror-like region, where the crack has grown slowly, surrounded by a region which is rough and coarse where the crack has propagated at speed (Figure 7.6).

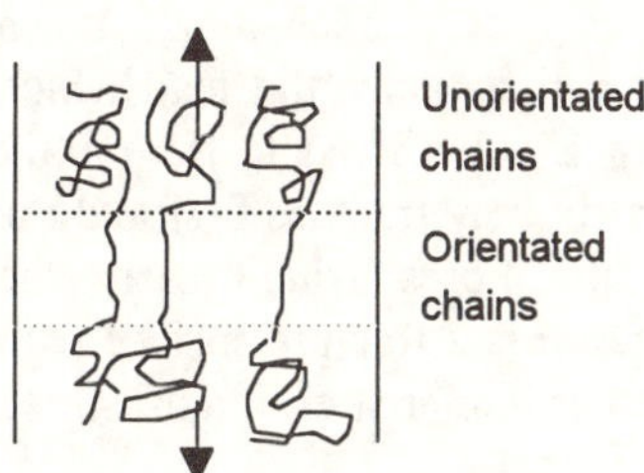

Figure 7.7 *Chain orientation*

In an amorphous polymer the chains are arranged randomly with no orientation. When stress is applied it can cause localised chain slippage and an orientation of molecule chains (Figure 7.7) with the result that the applied stress causes small voids to form between the aligned molecules and fine cracks, termed *crazing*, are formed. This is what constitutes the mirror-like region. Because of the inherent weakness of the material in the crazed region it serves as a place for cracks to propagate from and cause the material to fracture. Initially the crack grows by the growth of the voids along the midpoint of the craze. These then coalesce to produce a crack which then travels through the material by the growth of voids ahead of the advancing crack tip. This part of the fracture surface shows as the rougher region.

With crystalline polymers, the application of stress results in the folded molecular chains becoming unfolded and aligned. The result is then considerable, permanent deformation and necking. Prior to the material yielding and necking starting, the material is quite likely to begin to show a cloudy appearance. This is due to small voids being produced within the material. Further stress causes these voids to coalesce to produce a crack which then travels through the material by the growth of voids ahead of the advancing crack tip. Figure 7.8 shows typical forms of stress–strain graphs for polymers showing this form of failure.

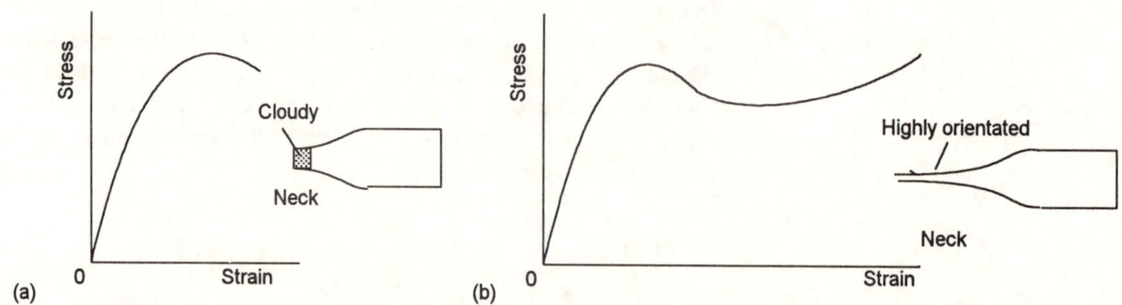

Figure 7.8 *Fracture with ductile polymeric materials: (a) unplasticised PVC, (b) polypropylene*

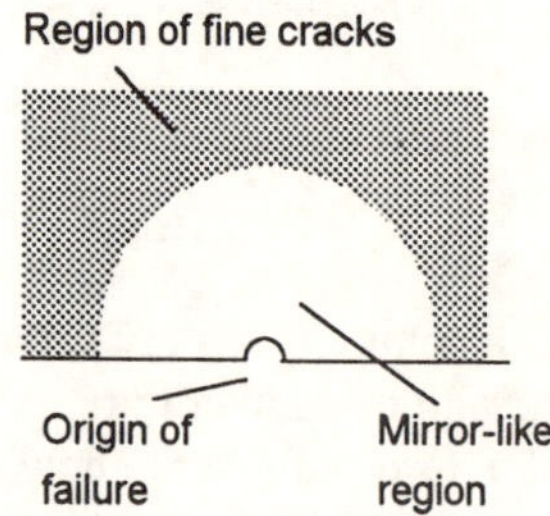

Figure 7.9 *Typical fracture surface for a ceramic*

5 *Failure with ceramics*
Ceramics are brittle materials, whether glassy or crystalline. Typically a fractured ceramic shows around the origin of the crack a mirror-like region bordered by a misty region containing numerous micro cracks (Figure 7.9). In some cases the mirror-like region may extend over the entire surface.

6 *Failure with composites*
The fracture surface appearances and mechanisms for composites depend on the fracture characteristics of the matrix and reinforcement materials and on the effectiveness of the bonding between the two. Thus, for example, for a glass-fibre reinforced polymer, depending on the strength of the bonds between fibres and polymer, the fibres may break first and then a crack propagate in shear along the fibre–matrix interface. Eventually the load which had been mainly carried by the fibres is transferred to the matrix which then fails. Alternatively the matrix may fracture first and the entire load is then transferred to the fibres which carry the increasing load until they break; the result is a fracture surface with lengths of fibre sticking out from it, rather like bristles of a brush.

Activity
Drinks are supplied in glass bottles, plastic bottles and aluminium cans. Compare the failures modes of these three materials.

7.3 Selection of materials and processes

The following are some case studies in the selection of materials and processes. There are no doubt alternative solutions to those suggested; in addition, there are likely to be many more factors involved before a selection is made.

7.3.1 Fizzy drink bottle

Consider the properties required of a bottle to contain, say, a fizzy drink:

1 High impact strength, not brittle, tough

2 Good barrier properties, i.e. the drink must not seep out through the container wall or lose its fizz due to a loss of carbon dioxide pressure

3 A material which will not taint the drink

4 Relatively stiff so that the bottle retains its shape

5 Able to withstand the pressure due to the carbon dioxide without deforming

6 Transparent and clear

7 Light weight

8 Cheap since the bottle is not intended to be reused (unlike milk bottles which are intended to be reused a number of times)

9 Large numbers required

Another factor determining the material that can be used is the processing method. Large numbers are required and so a processing method is required that has this capability. The product is hollow and so the obvious choice is blow moulding, assuming a polymer is to be chosen since a cheap, non-reusable, light weight product is required and this rules out glass.

The blow moulding requirement means that the material must be a thermoplastic. With regard to their impact properties, polymers can be grouped into three categories with the middle class sometimes subdivided into two. Category 1 are those polymeric materials which are brittle, even when unnotched; category 2 are those which are tough when unnotched but brittle when notched; category 3 are those which are tough under all conditions. Table 7.1 indicates the materials in each category. With regard to impact properties the choice of material is thus restricted to those in categories 2 and 3. Hence acrylics, though clear and transparent, are not suitable materials. Of those materials in categories 2 and 3 that are clear and transparent, the choice becomes low-density polyethylene, PVC or polyethylene terephthalate.

Table 7.1 *Impact properties of polymeric materials*

1. Brittle	2. Tough but brittle when notched	3. Tough
Acrylics	Polypropylene	Wet nylon
Glass-filled nylon	Cellulosics	Low-density polyethylene
Polystyrene	PVC	ABS (some forms)
	Dry nylon	Polycarbonate (some forms)
	Acetals	PTFE
	High-density polystyrene	
	ABS (some forms)	
	Polyethylene terephthalate	
	Polycarbonate (some forms)	

The bottles are to contain fizzy drinks, i.e. liquids containing carbon dioxide under pressure. Table 7.2 gives data concerning the permeability to water and carbon dioxide of the three materials. Despite being the tougher material, low-density polyethylene would not be a suitable material for a fizzy drink container because of its high permeability to carbon dioxide. The drink would not retain its fizz. Polyethylene terephthalate appears to be the best choice. This material is widely used for carbonated drink containers; PVC is used for non-carbonated drinks, e.g. wine.

Table 7.2 *Permeabilities to water and carbon dioxide*

Polymer	Permeability 10^{-8} mol mN^{-1} s^{-1}	
	To water	To carbon dioxide
Low-density polyethylene	30	5700
PVC	40	98
Polyethylene terephthalate	60	30

7.3.2 Car bodywork

The functions required of car bodywork are that it protects the engine and the car occupants from the weather and provides a pleasing appearance at not too high a cost. The properties required of the material used for the bodywork of cars include:

1 It can be formed to the shapes required

2 It has a smooth and shiny surface

3 Corrosion is not too significant

4 It is not brittle, being sufficiently tough to withstand small knocks, and relatively stiff

5 It is cheap, taking into account the costs of raw materials, processing and finishing

6 The material must be capable of being used in a process which is economic for large numbers

In considering metals, processing will be key factor in determining the material to be used. The shapes required and the fact that sheet material is required, together with the need for mass production, would suggest forming from sheet. Hot forming does present the problem of an unacceptable surface finish and so a material has to be chosen which allows for cold forming. This means a highly ductile material. Possibilities would be low-carbon steels or aluminium alloys. Table 7.3 gives the percentage elongations for some possible materials in the annealed state.

Table 7.3 *Ductilities of carbon steel and aluminium alloys*

Material	Percentage elongation
0.1% carbon steel	42
0.2% carbon steel	37
0.3% carbon steel	32
1.25% Mn, aluminium alloy	30
2.24% Mn, aluminium alloy	22

From the data in Table 7.3 it can be established that carbon steels and aluminium alloys could be used, both having enough ductility to enable sheet to be formed. In addition, both are reasonably tough. Aluminium alloys have the advantage of lower densities and so could lead to lower weight cars. The carbon steels do, however, have the advantages of work hardening more rapidly than the aluminium alloys. A material that work hardens rapidly is less likely to form a neck. In the case of cold forming of sheet, this would show as a thin region in the formed sheet which would clearly not be desirable. The great advantages of carbon steel, outweighing all other considerations, is that it is much cheaper. Thus the optimum material is a low-carbon steel; in practice a steel with less than 1% carbon is used.

Polymeric materials could be used for car bodywork. The problem with such materials is obtaining enough stiffness. One way of overcoming this is to form a composite material with glass fibre mat or cloth in a matrix of a thermoset. Unfortunately such a process of building up bodywork is essentially a manual rather than a machine process and so is slow. While it can be used for one-off bodies it is not suitable for mass production. Another possibility is to form glass-reinforced panels by hot pressing from sheet-moulding compound and then fitting the panels to a steel frame. This does enable a mass production method to be used. Such a method has been used for lorry cabs. Another possibility is to produce a sandwich type of composite for panels. This could be a foamed plastic between plastic or metal sheets. While such composite materials can give an appropriate stiffness, the costs tend to be higher than using steel.

7.3.3 Small components for toys

Consider small components such as the wheels for, say, a small model toy car for use by a small child. The functions required of the wheels are that they are safe and rotate on their axles. The materials thus need to be non-toxic, reasonably tough, not easily deformed by knocks, not brittle and cheap. Before considering possible materials, there is a British Standard, BS 5665, which should be consulted. This specifies, in Part 1, material, construction and design requirements for toys, methods of test for certain properties, and requirements for packaging and marketing. Part 2 specifies categories of flammable materials not to be used in the manufacture of toys. Part 3 gives the requirements and methods of test for migration of antimony, arsenic, barium, cadmium, chromium, lead, mercury and selenium from toy materials.

The products are required to be cheap when produced in relatively large quantities and the products themselves are rather small. In the case of metals, a possible process is die casting. Though the initial cost of the die is high, a large number of components can be produced from one die and so the cost per component becomes relatively low. In the case of polymers, a possible process is injection moulding. This also has a high die cost but large numbers of components can be produced from one die and hence the cost per component can be low. Both processes give a good surface finish and good dimensional accuracy.

In the case of metals, die casting limits the choice to those with relatively low melting points, i.e. aluminium, magnesium, zinc, lead and tin alloys. Table 7.4 shows relevant properties of these materials.

Table 7.4 *Die casting alloys at 20°C*

Alloy	Density Mg/m^3	Melting pt. °C	Strength MPa
Aluminium	2.7	600	150
Lead	11.3	320	20
Magnesium	1.8	520	150
Tin	7.3	230	12
Zinc	6.7	380	280

Safety considerations (see the British Standard) rule lead out. Aluminium, magnesium and zinc are comparable in cost, with zinc tending to have the lower cost per unit weight. Tin is more expensive than these alloys. Zinc has a lower melting point than aluminium or magnesium and in the as-cast condition has the highest tensile strength. Thus zinc would seem to be the best metal choice for the product. Zinc is very widely used for die casting. Its low melting point and excellent fluidity make it one of the easiest metals to cast. Small parts of complex shape and thin wall sections can be produced. Zinc alloys have relatively good mechanical properties and can be electroplated.

Polymers are a possible alternative to metals. The forming method capable of producing such items in quantity is injection moulding. The materials used are restricted to thermoplastics. The choice is then polymers which are relatively stiff. Table 7.5 shows how the properties of possible polymers compare with those of zinc. The mechanical properties of the zinc alloy are superior to those of thermoplastics, it having higher strength, higher tensile modulus, and being tougher, and more resistant to fatigue and creep. Where light weight is required, then polymers have the advantage, having densities of the order of one-sixth that of zinc. Where coloured surfaces are required then polymers have the advantage since pigments can be incorporated in the polymer mix. However, if electro-plating is required, then zinc has the advantage. On cost per unit weight zinc is cheaper; however, the interest is likely to be cost per unit volume.

Table 7.5 *Comparison of zinc and thermoplastics at 20°C*

Material	Relative cost/m³	Density Mg/m³	Strength MPa	Cost/unit strength	Modulus GPa	Cost/unit stiffness	Notch impact strength kJ/m²	Cost/unit impact strength
ABS	1	1.02–1.07	50	0.020	2.3	0.43	7	0.14
Nylon 6	2	1.13–1.14	60	0.033	3.2	0.63	3	0.67
Polycarbonate	2	1.2	65	0.031	2.3	0.87	30	0.07
Zinc alloy	3	6.7	280	0.011	103	0.03	55	0.05

Activity
Select a toy, consider the materials used in its contruction and produce reasoned arguments for the materials and processes used.

7.3.4 Tennis rackets

The function of a tennis racket is to transmit power from the arm of the player to a tennis ball. The requirements for the frame and handle of a racket are:

1 High strength

2 High stiffness

3 Low weight
The requirement for high strength and low weight can be translated into a requirement for a high value of strength/density, i.e. specific strength. Similarly the requirement for high stiffness and low weight into a requirement for a high value of modulus/density, i.e. specific modulus

4 Tough and able to withstand impact loading

5 The ability to damp out vibrations
When the ball hits the strings, the impact leads to vibrations of the racket. These are then transmitted through the frame of the racket to the arm of the player. If these vibrations are not reduced in amplitude in this transmission, the elbow of the player can suffer some damage, known as tennis elbow. The elbow joint does not like being vibrated.

6 Durable and does not creep or warp as a result of exposure to temperature or humidity changes

7 Can be processed into the required shape

8 Cost will be a factor when considering tennis rackets for the general population but less a requirement for rackets for professional tennis players.

Possibilities would seem to be wood, metals and composites. Table 7.6 shows typical values for some possible materials.

Table 7.6 *Materials for tennis rackets*

Material	Specific strength MPa/Mg m^{-3}	Specific stiffness GPa/Mg m^{-3}	Relative toughness	Relative vibration damping	Relative cost
Woods					
Ash	107	20	Good	Good	Low
Hickory	105	21	Good	Good	Low
Aluminium alloys					
Al–Cu alloy, precipitation hardened	15	25	Good	Poor	Medium
Al–Mg alloy, annealed	54	25	Good	Poor	Medium
Steels					
Mn steel, quenched and tempered	90	27	Good	Poor	Medium
Ni–Cr–Mo steel, quenched and tempered	115	27	Good	Poor	Medium
Composites					
Epoxy + 60% carbon	890	90	Medium	Medium	High
Epoxy + 70% glass	750	25	Medium	Medium	High

Wood has the advantages that it is tough, has good specific strength, good damping properties for vibrations and is cheap. The specific stiffness could be better. Warping could be a problem. However, this can be overcome by using laminated wood, i.e. several pieces of wood with their fibres in different directions bonded together to give a laminate. This combining together of pieces of wood also gives a method by which the shape of the racket can be obtained.

Aluminium alloys have the advantages of toughness and good specific stiffness. They are, however, more expensive than wood. Another problem is that they have very poor vibration damping. Aluminium can be protected against corrosion attack by damp environments by anodising. An aluminium racket could be made by bending extruded hollow sections into the required shape.

Steels can give high specific strengths and high specific stiffness. The steels with these high strengths are likely to be comparable in price with the aluminium alloys. Problems are, however, the very poor vibration damping and the poor corrosion resistance in a damp environment. A steel racket could be made by bending extruded hollow sections into the required shape.

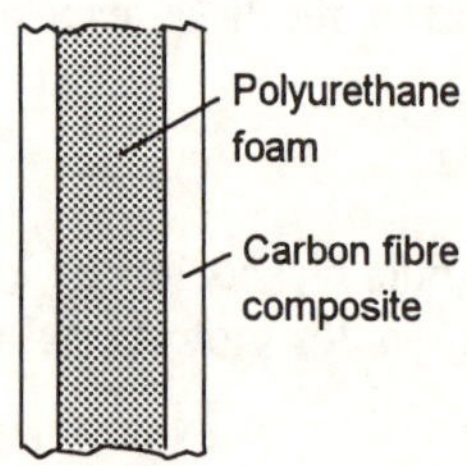

Figure 7.10 *Composite frame*

Composite materials can be made which have the advantages of very high specific strengths, very high specific stiffnesses, reasonable vibration damping and tolerable toughness. The major problem, however, is the high cost of such materials. A composite racket can be made by injecting a melt of a polymer containing carbon fibres into a racket-shaped mould. This would give a racket with a solid composite for the frame and handle (Figure 7.10). The procedure that can then be adopted to improve the properties is, while the racket is still in the mould and only the outer skin of the composite has solidified, to pour out the liquid core so that when the racket solidifies there is a hollow tube. The tube can then be filled with a polyurethane foam. This improves the vibration damping of the racket.

In comparing the above, the composite material racket gives the best properties but is considerably more expensive than the others. It thus is more likely to be used by the professional tennis player. For cheapness and properties, wood is probably the next best material, followed by aluminium alloys with steel being the worst.

7.3.5 Electrical resistors

Consider the requirements for resistors for general use in electrical and electronic circuits. They are required to obey Ohm's law and have resistances that do not markedly change when the temperature changes. This would suggest metals as the required material; the resistance of semiconductors changes much more than metals when the temperature changes. The values of resistance frequently required are in the range 1 kΩ to 10 MΩ. But metals tend to have resistivities ρ of the order of 10^{-6} Ω m. Consider therefore the problem of using such a material for a resistance of 1 kΩ. Suppose we use wire with a diameter of 1 mm. The length L of wire required will be given by:

$$\rho = 10^{-6} = \frac{RA}{L} = \frac{1000 \times \frac{1}{4}\pi \times 0.001^2}{L}$$

Hence the length required is 785 m. This is not practical. A more practical length, for a reasonably compact resistor, would be about 100 mm. This is only possible with such a resistivity if the cross-sectional area is 10^{-10} m^2. This would be a rectangular strip 0.01 mm by 0.01 mm or wire with a diameter of about 0.011 mm. How then can we produce such resistors?

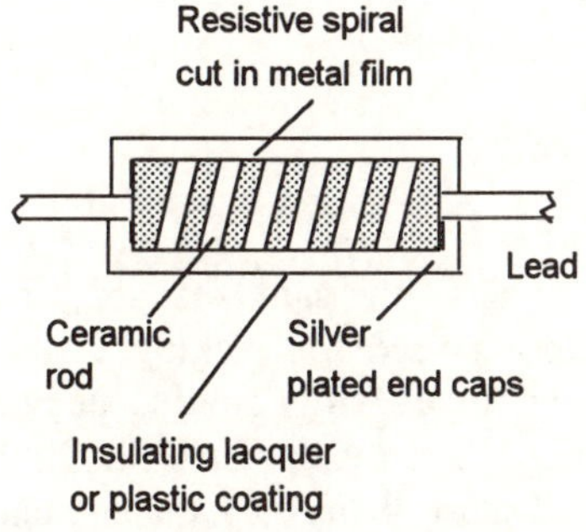

Figure 7.11 *Thin film resistor*

One possible method by which we can use metals is to deposit very thin layers on an insulating substrate. This can then give a very thin thickness of metal. Thin films of thickness about 10 nm (1 nm = 10^{-9} m) are used. We can then make this into a reasonable length by etching a suitable pattern in the metal. Thus a spiral groove might be cut through the metal deposited on a cylindrical substrate (Figure 7.11). The resistance value can be adjusted to some required value by stopping the groove cutting when that value is obtained. Nickel–chromium alloys (nichrome) are widely used for resistors manufactured in this way.

Another alternative is to mix a conductive powder with an insulator and organic solvent, the resulting mixture then being spread over an insulating substrate as a film about 10 μm thick. The conductive powders used are highly conductive oxides such as PdO or RuO_2. They have resistivities of about 10^{-6} Ω m and behave as metals. The mixture is fired so that organic solvents evaporate and the insulator and conductive material are left bonded to the substrate. The dispersed conductive particles are considered to form convoluted chains through the insulator. Thus we effectively end up with a number of exceedingly small cross-section conductors. The resistance value for the resistor is then determined by the concentration of the conductive powder in the insulator.

There is an alternative to using a metal and that is to use carbon in the form of graphite. Diamond is a crystal structure based solely on carbon atoms, each atom being bound by strong covalent bonds to four other carbon atoms. The result is a strong three-dimensional structure that is an electrical insulator because there are no free electrons. Diamond is also very hard. Graphite, however, is a very soft material. It is the material used as the lead in pencils. Graphite, like diamond, consists only of carbon atoms. However, the way in which the carbon atoms are arranged in the solid is quite different (Figure 7.12). It can be considered to be a 'layered' structure. The atoms are strongly bonded together with covalent bonds in two-dimensional layers, with only very weak bonds, van der Waals bonds, between the atoms in different layers. As a result, there are free electrons between the layers and the material conducts electricity. At room temperature, the resistivity of diamond is about 10^{10} Ω m, while that of graphite is about 10^{-6} Ω m.

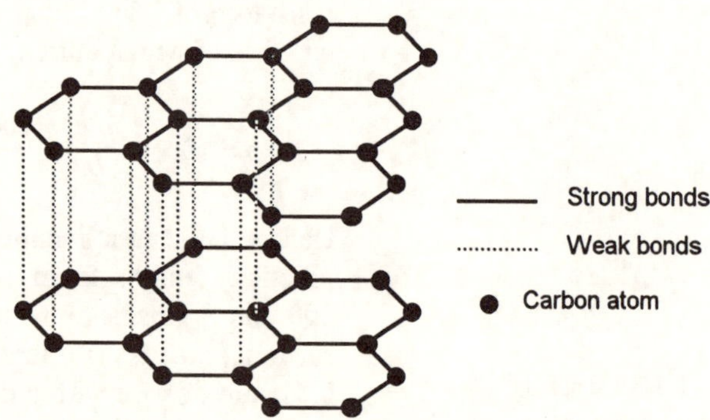

Figure 7.12 *Graphite*

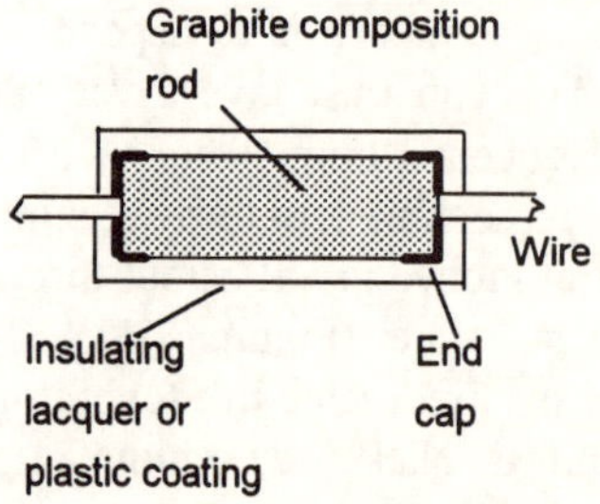

Figure 7.13 *Graphite composition resistor*

Resistors using graphite all tend to be about the same size, but can have a wide range of values. The process used for making carbon resistors involves powdered graphite being mixed with naphthalene and an insulating filler such as china clay. The mix is then pressed into little cylinders and then fired. This gets rid of most of the naphthalene and leaves a porous graphite structure. The amount of naphthalene, and hence the degree of porosity, determine the resistance of the cylinder. Figure 7.13 shows the form such a resistor might take.

7.3.6 Car engine pistons

Figure 7.14 shows the basic form of an internal combustion engine and the piston. Fuel enters the cylinder and is ignited by a spark at the spark plug. The piston then has the function of retaining the hot expanding gas on one side, moving as a result of the expansion and communicating the movement via the piston rod to the crankshaft; it then moves back up the

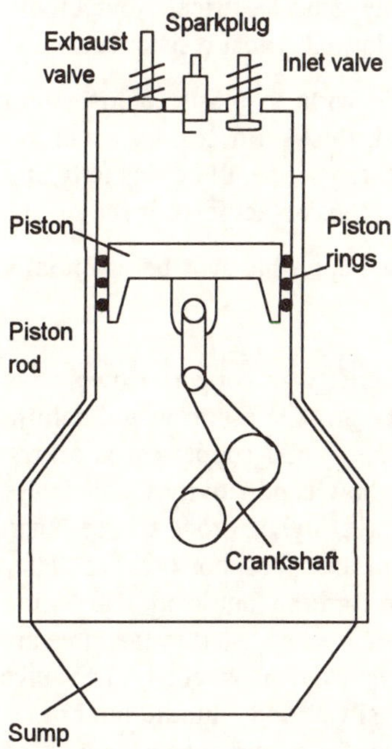

Figure 7.14 *Internal combustion engine*

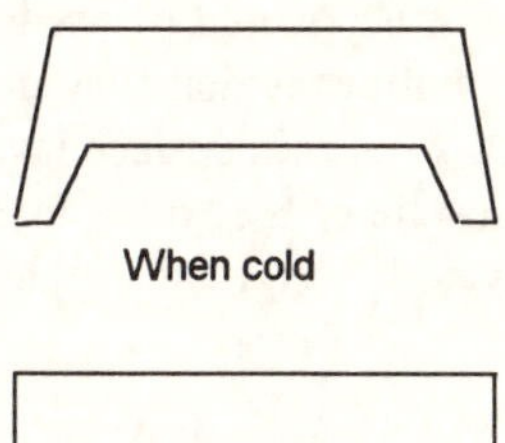

Figure 7.15 *Piston design*

cylinder and expels the products of the combustion through the exhaust valve.

Pistons are subject to high temperatures on one side and relatively cold on the other. They thus have the problem of accommodating thermal expansion since they must be a reasonably tight fit in the cylinder but not expand so much that they become a tight fit. The upper part of a piston is fitted with piston rings to help retain the combustion gases in the upper part of the cylinder and also to take the wear resulting from sliding up and down the cylinder wall and avoid the piston rubbing against the wall and wearing.

Earlier low speed engines had pistons of cast iron to match the material used for the cylinder. With increasing engine speeds, modern pistons are made from an aluminium alloy, e.g. the aluminium casting alloy LM13. Such an alloy has a high thermal conductivity and also has the great benefit of a low density, so allowing lighter weight pistons. The high thermal conductivity enables heat to be more rapidly conducted away and so results in the piston running at a lower temperature than otherwise would be the case. The material used must retain its properties at temperatures up to about 300°C. Thermal expansion is, however, a problem. This is particularly the case when they are in a cylinder made of cast iron since cast iron has a different coefficient of thermal expansion to that of an aluminium alloy; Table 7.7 gives thermal data for both materials. In the absence of special design features, the difference in expansion could result in seizure when hot (leaving enough room for the expansion would mean too loose a fitting piston when cold and result in 'piston slap'). Such design features include the piston being tapered (Figure 7.15) so that there is additional clearance when cold near the top; it is this part which attains the highest temperature when hot and thus expands more than the cooler lower part of the piston. Casting or forging can be used to manufacture the pistons.

Table 7.7 *Thermal properties*

Material	Density Mg/m^3	Thermal conductivity $W\ m^{-1}K^{-1}$	Coefficient of thermal expansion $10^{-6}/°K$
Grey cast iron	7.2	44–53	11
Aluminium alloy LM13	2.7	117	19

7.3.7 Electrical switch contacts

Electrical switch contacts, e.g. as in Figure 7.16, and sockets into which electronic device pins are push-fits need:

1 To make contact which has a low electrical contact resistance

2 To maintain a low contact resistance after repeated use and exposure to air and hence oxidation

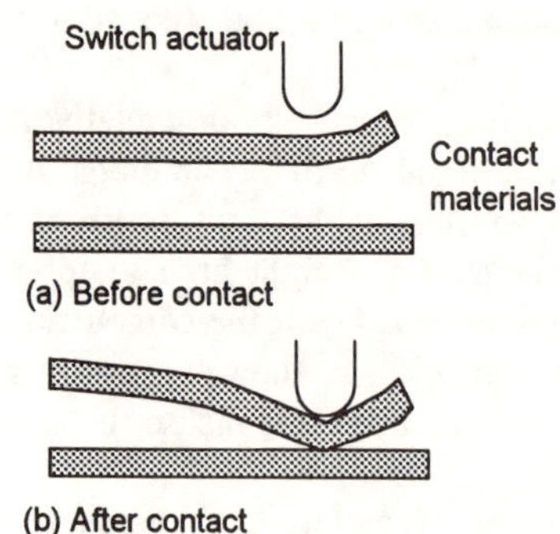

Figure 7.16 *Electrical contact*

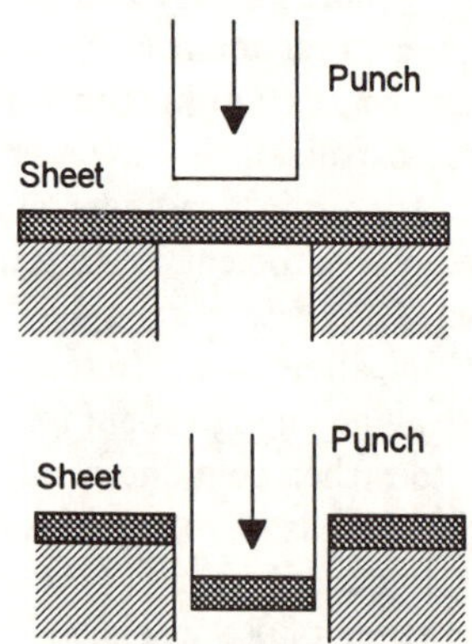

Figure 7.17 *Blanking*

3 The material needs to have reasonably good electrical conductivity in order that the strips used will have low electrical resistance

4 The material is required to be elastic with no plastic deformation over the range of stresses experienced; this is a necessary condition if the contacts are to be pressed together as a result of elasticity and if the contacts are to spring back when the contact force is removed

5 A forming process is likely to be used for what will be essentially strips of metal

The obvious material for good conductivity is copper. However, a relatively pure copper is ductile with a low proof stress so would deform very easily. Table 7.8 shows data for coppers and copper-based alloys. All the metals in the table have an acceptably conductivity for their use as strips. Brasses and the nickel silver have higher proof stresses than coppers but even better are the aluminium and phosphor bronzes. They have sufficiently ductility when soft for forming but work hardening gives a material with good elasticity. All the metals listed in the table are suitable for forming; the coppers, red brass and phosphor bronze have very good formability, with the nickel silver and aluminium bronze having good formability. A possible process for strip contacts is blanking from sheet; this process involves a punch which presses down on the sheet and punches the required shapes out (Figure 7.17).

The requirement for low contact resistance, with this being maintained over time and after repeated contacts having being made, requires testing of the materials in the conditions they would be used in. A problem with coppers and many copper-based alloys is that they have excellent corrosion reistance but that this results from thin surface layers of oxide being formed on exposure to the atmosphere. It is these layers which prevent further corrosion. However, they do have a relatively high resistance and so will affect contact resistance.

Table 7.8 *Data for coppers and copper-based alloys*

Material	0.2% proof stress MPa	Tensile strength MPa	% elongation	Electrical conductivity % IACS
Electrolytic tough-pitch h.c. copper C101	50–320	220–400	50–4	101.5–100
Oxygen-free h.c. copper C103	50–320	220–400	60–6	101.5–100
Red brass CZ102	120–140	310–330	40	32
Common brass CZ108, soft sheet	130	340	56	27
hard sheet	180	530	5	26
Aluminum brass CZ110, soft sheet	125	340	60	23
15% nickel silver NS105, soft sheet	130–180	390–420	55–45	7
hard sheet	370–600	490–710	15–2	7
95/5 Aluminium bronze CA101, soft sheet	370	140	65	15–18
hard sheet	650	540	15	15–18
3% Phosphor bronze PB101, soft strip	110	320	55	15–25
hard strip	450–460	580–590	8	15–25

In normal atmospheric conditions, copper acquires a thin surface film of oxide, this can markedly affect contact resistance. The brasses have good corrosion resistance but surface layers result in a significant contact resistance. The nickel silver has reasonable resistance to corrosion but significant contact resistance. The aluminium bronze has very good corrosion resistance and maintains a particularly low contact resistance. The phosphor bronze has a good corrosion resistance and maintains a low contact resistance. The phosphor bronze and the aluminium bronze are thus widely used for switch contacts.

Activity
Select a product and give explanations as to why you think the materials it is formed from were selected and what processes might have been used to make it.

Problems

1 Make reasoned proposals for materials and processes for:
(a) domestic window catches,
(b) structural I-beams for use in building construction,
(c) rainwater gutters and drainpipes,
(d) a domestic washing-up bowl,
(e) pipe through which sea water can be pumped,
(f) small fan in a vacuum cleaner,
(g) the lenses for the rear lights of cars,
(h) a camshaft for a car,
(i) casing for a hand-held power tool,
(j) the blades for a Flymo hover mower.

2 Investigate the materials used with the following products and give reasons why they might have been chosen in preference to others and the processes that might have been used:
(a) the casing for mains electric plugs,
(b) spades,
(c) domestic cold and hot water pipes,
(d) the casing for the body of a vacuum cleaner,
(e) joists to support floors in a small house.

3 For the following consult the British Standard specified and then present reasons for the choice of materials and processes for the specified products:
(a) BS 4654 Hooks for lifting freight containers,
(b) BS 3388 Forks, shovels and spades,
(c) BS 3441 Tanks for the transport of milk,
(d) BS 3531 Section 5.2 Screwdrivers,
(e) BS 4109 Copper wire for electrical purposes,
(f) BS 4344 Pulley blocks for use with natural and synthetic fibre ropes,
(g) BS 4637 Coiled springs,
(h) BS 6728 Hot water bottles made from PVC compounds,
(i) BS 7413/4 Unplasticised PVC for window frames.

Appendix A: Units

The International System (SI) of units has seven base units, these being:

Length	metre	m
Mass	kilogram	kg
Time	second	s
Electric current	ampere	A
Temperature	kelvin	K
Luminous intensity	candela	cd
Amount of substance	mole	mol

In addition there are two supplementary units, the radian and the steradian.

The SI units for other physical quantities are formed from the base units via the equation defining the quantity concerned. Thus, for example, volume is defined by the equation volume = length cubed, thus the unit of volume of the length unit cubed, i.e. m^3. Density is defined by the equation density = mass/volume and thus has the unit of mass divided by the unit of volume, i.e. kg/m^3. Velocity is defined by the equation velocity = change of displacement/time taken and thus has the unit of length divided by the unit of time, i.e. m/s. Acceleration is defined by the equation acceleration = change in velocity/time taken and thus has the unit of velocity divided by the unit of time, i.e. (m/s)/s or m/s^2.

Some of the derived units are given special names. Thus, for example, the unit of force is defined by the equation force = mass × acceleration and is thus kg m/s or kg m s^{-1}. This unit is given the name newton (N). Thus 1 N = 1 kg m s^{-1}. The unit of stress is given by the equation stress = force/area and thus has the derived unit of N/m^2. This unit is given the name pascal (Pa). Thus 1 Pa = 1 N/m^2.

Certain quantities are defined as the ratio of two comparable quantities. Thus, for example, strain is defined as change in length/length. It thus is expressed as a pure number with no units because the derived unit would be m/m.

Standard prefixes are used for multiples and submultiples of units, the SI preferred ones being multiples of 10^3. The following are the standard prefixes:

Thus, for example, 1000 N can be written as 1 kN, 1 000 000 Pa as 1 MPa, 1 000 000 000 Pa as 1 GPa, 0.001 m as 1 mm, and 0.000 001 A as 1 μA. Note that often the unit N/mm^2 is used for stress, 1 N/mm^2 is 1 MPa.

Multiplication factor	Prefix	
1 000 000 000 000 000 000 000 000 = 10^{24}	yotta	Y
1 000 000 000 000 000 000 000 = 10^{21}	zetta	Z
1 000 000 000 000 000 000 = 10^{18}	exa	E
1 000 000 000 000 000 = 10^{15}	peta	P
1 000 000 000 000 = 10^{12}	tera	T
1 000 000 000 = 10^{9}	giga	G
1 000 000 = 10^{6}	mega	M
1 000 = 10^{3}	kilo	k
100 = 10^{2}	hecto	h
10 = 10	deca	da
0.1 = 10^{-1}	deci	d
0.01 = 10^{-2}	centi	c
0.001 = 10^{-3}	milli	m
0.000 001 = 10^{-6}	micro	μ
0.000 000 001 = 10^{-9}	nano	n
0.000 000 000 001 = 10^{-12}	pico	p
0.000 000 000 000 001 = 10^{-15}	femto	f
0.000 000 000 000 000 001 = 10^{-18}	atto	a
0.000 000 000 000 000 000 001 = 10^{-21}	zepto	z
0.000 000 000 000 000 000 000 001 = 10^{-24}	yocto	y

For further information about SI units, including their definitions, the reader is referred to SI The International System of Units, edited by R.J. Bell (National Physical Laboratory, HMSO). This is the approved translation of the International Bureau of Weights and Measures publication.

Other units which the reader may come across are fps (foot-pound-second) units which still are often used in the USA. On that system the unit of length is the foot (ft), with 1 ft = 0.3048 m. The unit of mass is the pound (lb), with 1 lb = 0.4536 kg. The unit of time is the second, the same as the SI system. With this system the derived unit of force, which is given a special name, is the poundal (pdl), with 1 pdl = 0.1383 N. However, a more common unit of force is the pound force (lbf). This is the gravitational force acting on a mass of 1 lb and consequently, since the standard value of the acceleration due to gravity is 32.174 ft/s^2, then 1 lbf = 32.174 pdl = 4.448 N. The similar unit the kilogram force (kgf) is sometimes used. This is the gravitational force acting on a mass of 1 kg and consequently, since the standard value of the acceleration due to gravity is 9.806 65 m/s^2, then 1 kgf = 9.806 65 N. A unit often used for stress in the USA is the psi, or pound force per square inch. 1 psi = 6.895×10^3 Pa.

Appendix B: Terminology

The following are some of the terms commonly encountered in discussing materials for engineering.

Additives Plastics and rubbers almost invariably contain, in addition to the polymer or polymers, other materials, i.e. additives. These are added to modify the properties and cost of the material.

Ageing This term is used to describe a change in properties that occurs with certain metals due to precipitation occurring, there being no change in chemical composition.

Alloy This is a metal which is a mixture of two or more elements.

Amorphous An amorphous material is a non-crystalline material, i.e. it has a structure which is not orderly.

Annealing This involves heating to and holding at a temperature which is high enough for recrystallisation to occur and which results in a softened state for a material after a suitable rate of cooling, generally slowly. The purpose of annealing can be to facilitate cold working, improve machineability, improve mechanical properties, etc.

Anodising This term is used to describe the process, generally with aluminium, whereby a protective coating is produced on the surface of the metal by converting it to an oxide.

Austenite This term describe the structure of a face-centred cubic iron crystalline structure which has carbon atoms in the gaps in the face-centred iron.

Bend, angle of The results of a bend test on a materials are specified in terms of the angle through which the material can be bent without breaking. The greater the angle the more ductile the material.

Brinell number The Brinell number is the number given to a material as a result of a Brinell test and is a measure of the hardness of a material. The larger the number the harder the material.

Brittle failure With brittle failure a crack is initiated and propagates prior to any significant plastic deformation. The fracture surface of a metal with a brittle fracture is bright and granular due to the reflection of light from individual crystal surfaces. With polymeric materials the fracture surface may be smooth and glassy or somewhat splintered and irregular.

Brittle material A brittle material shows little plastic deformation before fracture. The material used for a china teacup is brittle. Thus because there is little plastic deformation before breaking, a broken teacup can be stuck back together again to give the cup the same size and shape as the original.

Carburizing This is a treatment which results in a hard surface layer being produced with ferrous alloys. The treatment involves heating the alloy in a carbon-rich atmosphere so that carbon diffuses into the surface layers, then quenching to convert the surface layers to martensite.

Case hardening This term is used to describe processes in which by changing the composition of surface layers of ferrous alloys a hardened surface layer can be produced.

Casting This is a manufacturing process which involves pouring liquid metal into a mould or, in the case of plastics, the mixing of the constituents in a mould.

Cementite This is a compound formed between iron and carbon, often referred to as iron carbide. It is a hard and brittle material.

Charpy test value The Charpy test is used to determine the response of a material to a high rate of loading and involves a test piece being struck a sudden blow. The results are expressed in terms of the amount of energy absorbed by the test piece when it breaks. The higher the test value the more ductile the material.

Cold working This is when a metal is subject to working at a temperature below its recrystallisation temperature.

Composite This is a material composed of two different materials bonded together in such a way that one serves as the matrix surrounding fibres or particles of the other.

Compressive strength The compressive strength is the maximum compressive stress a material can withstand before fracture.

Copolymer This is a polymeric material produced by combining two or more monomers in a single polymer chain.

Corrosion resistance This is the ability of a material to resist deterioration by reacting with its immediate environment. There are many forms of corrosion and so there is no unique way of specifying the corrosion resistance of a material.

Creep Creep is the continuing deformation of a material with the passage of time when it is subject to a constant stress. For a particular material the creep behaviour depends on both the temperature and the initial stress, the behaviour also depending on the material concerned.

Crystalline This term is used to describe a structure in which there is a regular, orderly, arrangement of atoms or molecules.

Damping capacity The damping capacity is an indicator of the ability of a material to suppress vibrations.

Density Density is mass per unit volume.

Dielectric strength The dielectric strength is a measure of the highest potential difference an insulating material can withstand without electric breakdown. It is the breakdown voltage divided by the thickness of the material.

Ductile failure With ductile failure there is a considerable amount of plastic deformation prior to failure. With metals the fracture shows a typical cone and cup formation and the fracture surfaces are rough and fibrous in appearance.

Ductile materials Ductile materials show a considerable amount of plastic deformation before breaking. Such materials have a large value of percentage elongation.

Elastic limit The elastic limit is the maximum force or stress at which on its removal the material returns to its original dimensions.

Electrical conductance This is the reciprocal of the electrical resistance and has the unit of the siemen (S). It is thus the current through a material divided by the voltage across it.

Electrical conductivity The electrical conductivity is defined by

$$\text{conductivity} = \frac{L}{RA}$$

where R is the resistance of a strip of the material of length L and cross-sectional area A. Conductivity has the unit of S/m. The IACS specification of conductivity is based on 100 per cent corresponding to the conductivity of annealed copper at 20°C, all other materials are then expressed as a percentage of this value.

Electrical resistance This is the voltage across a material divided by the current through it, the unit being the ohm (Ω).

Electrical resistivity The electrical resistivity is defined by

$$\text{resistivity} = \frac{RA}{L}$$

Resistivity has the unit Ω m.

Expansion, coefficient of linear The coefficient of linear expansion is a measure of the amount by which a unit length of a material will expand when the temperature rises by one degree. It is defined by

$$\text{coefficient} = \frac{\text{change in length}}{\text{length} \times \text{temp. change}}$$

It has the unit /°C or $°C^{-1}$ or /K or K^{-1}.

Expansivity, linear This is an alternative name for the coefficient of linear expansion.

Fatigue life The fatigue life is the number of stress cycles to cause failure.

Ferrite This term is usually used for a structure consisting of carbon atoms lodged in body-centred cubic iron. Ferrite is comparatively soft and ductile.

Fracture toughness The plane strain fracture toughness is an indicator of whether a crack will grow or not and thus is a measure of the toughness of a material when there is a crack present.

Full hard This term is used to describe the temper of alloys. It corresponds to the cold-worked condition beyond which the material can no longer be worked.

Grain This term is used for a crystalline region within a metal, i.e. a region of orderly packed atoms.

Half hard This term is used to describe the temper of alloys. It corresponds to the cold-worked condition half-way between soft and full hard.

Hardness The hardness of a material may be specified in terms of some standard test involving indentation, e.g. the Brinell, Vickers and Rockwell tests, or scratching of the surface of the material, the Moh test.

Heat treatment This term is used to describe the controlled heating and cooling of metals in the sold state for the purpose of altering their properties.

Hooke's law When a material obeys Hooke's law its extension is directly proportional to the applied stretching forces.

Hot working This is when a metal is subject to working at a temperature in excess of its recrystallisation temperature.

Impact properties See Charpy test value and Izod test value.

Izod test value The Izod test is used to determine the response of a material to a high rate of loading and involves a test piece being struck a sudden blow. The results are expressed in terms of the amount of energy absorbed by the test piece when it breaks. The higher the test value the more ductile the material.

Limit of proportionality Up to the limit of proportionality the extension is directly proportional to the applied stretching forces, i.e. the strain is proportional to the applied stress.

Malleability This describes the ability of metals to permit plastic deformation in compression without rupturing.

Martensite This is a general term used to describe a form of structure. In the case of ferrous alloys it is a structure produced when the rate of cooling from the austenitic state is too rapid to allow carbon atoms to diffuse out of the face-centred cubic form of austenite and produce the body-centred form of ferrite. The result is a highly strained hard structure.

Melting point This is the temperature at which a material changes from solid to liquid.

Moh scale This is a scale of hardness arrived at by considering the ease of scratching a material. It is a scale of 10, with the higher the number the harder the material.

Monomer This is the unit, or mer, consisting of a relatively few atoms which are joined together in large numbers to form a polymer.

Nitriding This is a treatment in which nitrogen diffuses into surface layers of a ferrous alloy and hard nitrides are produced, hence a hard surface layer.

Orientation A polymeric material is said to have an orientation, uniaxial or biaxial, if during the processing of the material the molecules become aligned in particular directions. The properties of the material in such directions are markedly different from those in other directions.

Pearlite This is a lamella structure of ferrite and cementite.

Percentage elongation The percentage elongation is a measure of the ductility of a material, the higher the percentage the greater the ductility. It is the change in length which has occurred during a tensile test to breaking expressed as a percentage of the original length.

$$\%\text{elongation} = \frac{\text{final} - \text{initial lengths}}{\text{initial length}} \times 100$$

Percentage reduction in area This is a measure of the ductility of a material and is the change in cross-sectional area which has occurred during a tensile test to breaking expressed as a percentage of the original cross-sectional area.

Precipitation hardening This is a heat treatment process which results in a precipitate being produced in such a way that a harder material is produced.

Proof stress The 0.2% proof stress is defined as that stress which results in a 0.2% offset, i.e. the stress given by a line drawn on the stress-strain graph parallel to the linear part of the graph and passing through the 0.2% strain value. The 0.1% proof stress is similarly defined. Proof stresses are quoted when a material has no well defined yield point.

Quenching This is the method used to produce rapid cooling. In the case of ferrous alloys it involves cooling from the austenitic state by immersion in cold water or an oil bath.

Recovery This term is used in the treatment involving the heating of a metal so as to reduce internal stresses.

Recrystallization This is generally used to describe the process whereby a new, strain free grain structure is produced from that existing in a cold-worked metal by heating.

Resilience This term is used with elastomers to give a measure of the 'elasticity' of a material. A high resilience material will suffer elastic collisions when a high percentage of the kinetic energy before the collision is returned to the object after the collision. A less resilient material would loose more kinetic energy in the collision.

Rockwell test value The Rockwell test is used to give a value for the hardness of a material. There are a number of Rockwell scales and thus the scale being used must be quoted with all test results.

Ruling section The limiting ruling section is the maximum diameter of round bar at the centre of which the specified properties may be obtained.

Secant modulus For many polymeric materials there is no linear part of the stress-strain graph and thus a tensile modulus cannot be quoted. In such cases the secant modulus is used. It is the stress at a value of 0.2% strain divided by that strain.

Shear When a material is loaded in such a way that one layer of the material is made to slide over an adjacent layer then the material is said to be in shear.

Shear strength The shear strength is the shear stress required to produce fracture.

Sintering This is the process by which powders are bonded by molecular or atomic attraction as a result of heating to a temperature below the melting points of the constituent powders.

Solution treatment This heat treatment involves heating an alloy to a suitable temperature, holding at that temperature long enough for one or more constituent elements to enter into the crystalline structure, and then cooling rapidly enough for these to remain in solid solution.

Specific gravity The specific gravity of a material is the ratio of its density compared with that of water.

Specific heat capacity The amount by which the temperature rises for a material when there is a heat input depends on its specific heat capacity. The higher the specific heat capacity the smaller the rise in temperature per unit mass for a given heat input.

$$\text{specific heat capacity} = \frac{\text{heat input}}{\text{mass} \times \text{temp. change}}$$

Specific heat capacity has the unit J kg^{-1} K^{-1}.

Specific stiffness This is the modulus of elasticity divided by the density.

Specific strength This is the strength divided by the density.

Stiffness The property is described by the modulus of elasticity.

Strain The engineering strain is defined as the ratio (change in length)/(original length) when a material is subject to tensile or compressive forces. Shear strain is the ratio (amount by which one layer slides over another)/(separation of the layers). Because it is a ratio strain it has no units, though it is often expressed as a percentage. Shear strain is usually quoted as an angle in radians.

Strength See Compressive Strength, Shear strength and Tensile strength.

Stress In engineering, tensile and compressive stress is defined as (force)/(initial cross-sectional area); true stress is (force)/(cross-sectional area at that force). Shear stress is the (shear force)/(area resisting shear). Stress has the unit Pa (pascal) with 1 Pa = 1 N/m^2 or 1 N m^{-2}.

Stress relieving This is a treatment to reduce residual stresses by heating the material to a suitable temperature, followed by slow cooling.

Stress-strain graph The stress-strain graph is usually drawn using the engineering stress (see Stress) and engineering strain (see Strain).

Surface hardening This is a general term used to describe a range of processes by which the surface of a ferrous alloy is made harder than its core.

Temper This term is used with non-ferrous alloys as an indication of the degree of hardness/strength, with expressions such as hard, half-hard, three-quarters hard being used.

Tempering This is the heating of a previously quenched material to produce an increase in ductility.

Tensile modulus The tensile modulus, or Young's modulus, is the slope of the stress-strain graph over its initial straight line region.

Tensile strength This is defined as the maximum tensile stress a material can withstand before breaking.

Thermal conductivity The rate at which energy is transmitted as heat through a material depends on a property called the thermal conductivity. The higher the thermal conductivity the greater the rate at which heat is conducted. Thermal conductivity is defined by

$$\text{thermal conductivity} = \frac{\text{rate of transfer of heat}}{\text{area} \times \text{temp. gradient}}$$

Thermal conductivity has the unit W m^{-2} K^{-1}.

Thermal expansivity See Expansion, coefficient of linear.

Toughness This property describes the ability of a material to absorb energy and deform plastically without fracturing. It is usually measured with the Izod test or the Charpy test. Another form of measure is the fracture toughness. See Fracture toughness.

Transition temperature The transition temperature is the temperature at which a material changes from giving a ductile failure to giving a brittle failure.

Vickers' test results The Vickers test is used to give measure of the hardness of a material; the higher the Vickers hardness number the greater the hardness.

Water absorption This is the percentage gain in weight of a polymeric material after immersion in water for a specified amount of time under controlled conditions.

Wear resistance This is a subjective comparison of the wear resistance of materials. There is no standard test.

Work hardening This is the hardening of a material produced as a consequence of working subjecting it to plastic deformation at temperatures below those of recrystallization.

Yield point For many metals, when the stretching forces applied to a test piece are steadily increased a point is reached when the extension is no longer proportional to the applied forces and the extension increases more rapidly than the force until a maximum force is reached. This is called the upper yield point. The force then drops to a value called the lower yield point before increasing again as the extension is continued.

Young's modulus See Tensile modulus.

Appendix C: Materials

The following lists commonly encountered engineering metals, polymers and ceramics with their main characteristics and properties.

Engineering metals

The following is an alphabetical listing of metals, each being listed according to the main alloying element, with their key characteristics. It is not a complete list of all metals, just those commonly encountered in engineering, with details of some of their properties.

Aluminium Used in commercially pure form and alloyed with copper, manganese, silicon, magnesium, tin and zinc. Alloys exist in two groups: casting alloys and wrought alloys (Tables C1, C2, C3). Some alloys can be heat treated. Aluminium and its alloys have a low density, high electrical and thermal conductivity and excellent corrosion resistance. Tensile strength tends to be of the order of 150 to 400 MPa with the tensile modulus about 70 GPa. There is a high strength to weight ratio. They are used for such applications as engine parts, car trims, aircraft structures, fan blades, cooking utensils, metal boxes and chemical equipment.

Chromium Chromium is mainly used as an alloying element in stainless steels, heat resistant alloys and high strength alloy steels. It is generally used in these for the corrosion and oxidation resistance it confers on the alloys.

Table C1 *Properties of wrought non-heat treatable aluminium alloys*

Alloy	Composition %	Temper	0.2% proof stress MPa	Tensile strength MPa	Percentage elongation
3003	1–1.5 Mn	Soft	25	110	30
		Work hardened	185	200	4
3004	1–1.5 Mn, 0.8–1.3 Mg	Soft	70	180	20
		Cold worked	250	280	5
5050	0.2 Mn, 0.5–1.1 Mg	Soft	55	145	24
		Cold worked	200	220	6
5182	0.35 Mn, 4.5 Mg	Soft	140	275	25
		Cold worked	285	340	10
5454	0.5–1 Mn, 2.4–3.0 Mg	Soft	120	250	22
		Cold worked	240	305	10

Table C2 *Properties of wrought, heat treatable, aluminium alloys*

Alloy	Composition %	Temper	0.2% proof stress MPa	Tensile strength MPa	Percentage elongation
2011	5–6 Cu, 0.4 Si, 0.3 Zn	TF	295	390	17
2014	3.9–5.0 Cu, 0.2–0.8 Mg, 0.4–1.2 Mn, 0.5–1.2 Si, 0.25 Zn	TF	410	480	13
6061	0.15–0.4 Cu, 0.8–1.2 Mg, 0.15 Mn, 0.4–0.8 Si, 0.25 Zn	TF	275	310	12
7001	1.6–2.6 Cu, 2.6–3.4 Mg, 6.8–8.0 Zn, 0.2 Mn, 0.35 Si	TF	625	675	9

TF = solution treated and artificially aged

Table C3 *Properties of commonly used aluminium casting alloys*

Alloy	Composition %	Temper	Casting process	0.2% proof stress MPa	Tensile strength MPa	Percentage elongation
LM6	10–13 Si	M	S	65	185	8
			GD	90	205	9
			PD	130	250	2.5
LM2	9–11.5 Si, 0.7–1.5 Cu	M	PD	150	250	3
LM5	3–6 Mg, 0.3–0.7 Mn	M	S	80	170	5
		M	GD	80	230	10

S = sand cast, GD = gravity die cast, PD = pressure die cast; M = as manufactured

Cobalt Cobalt is widely used as an alloy for magnets, typically 5–35% cobalt with 14–30% nickel, and 6–13% aluminium. Cobalt is also used for alloys which have high strength and hardness at room and high temperatures. These are often referred to as Stellites. Cobalt is also used as an alloying element in steels.

Copper Copper is very widely used in the commercially pure form (Table C4) and alloyed in the form of brasses, bronzes, cupro-nickels, and nickel silvers. Brasses are copper–zinc alloys containing up to 43% zinc (Table C5) and are used for decorative and architectural items, coins, medals, fasteners, locks, hinges, pins and rivets. Bronzes are copper-tin alloys (Table C6) and are used for screws, bolts, rivets, springs, clips, bellows and diaphragms. Copper–aluminium alloys are referred to as aluminium bronzes (Table C7) and are used for nuts, bolts, bearings and heat-exchanger tubes. Copper–beryllium alloys are referred to as beryllium bronzes (Table C8) and are used for springs, clips and fasteners. Copper-silicon alloys as referred to as silicon bronzes (Table C9) and are used for chemical and marine plant items. Cupronickels are referred to as copper-nickel alloys and are used for coins, medals and where high corrosion resistance is required to sea water. If zinc is added to copper-nickel alloys, the alloy is termed a nickel silver (Table C10) and is used for clock and watch components, rivets, screws, clips and decorative items. Copper and its alloys have good corrosion resistance, high electrical and thermal conductivity, good machinability, can be joined by soldering, brazing and welding, and generally have good properties at low temperatures. The alloys have tensile strengths from about 180 to 300 MPa and a tensile modulus about 20 to 28 GPa.

Table C4 *Properties of coppers*

Ref.	Copper	Composition %	Condition	Tensile strength MPa	Percentage elongation	Brinell hardness	Electrical conductivity IACS %
C101	Electrolytic tough pitch hc	Cu 99.90 min., 0.05 O	Annealed	220	50	45	101.5–100
			Hard	400	4	115	
C103	Oxygen-free hc	Cu 99.95 min.	Annealed	220	60	45	101.5–100
			Hard	400	6	115	
C108	Cadmium copper	Cu 99.0, 1 Cd	Annealed	280	45	95	75–92
			Hard	700	4	145	
C110	Oxygen-free hc copper	Cu 99.99 min.	Annealed	220	60	45	101.5–100
			Hard	400	6	115	

The electrical conductivities have been expressed on the IACS scale, the value of 100% on this scale corresponding to the electrical conductivity of annealed copper at 20°C with conductivity 5.800×10^7 S/m.

Table C5 *Properties of brasses*

Alloy	Name	Composition %	Condition	Tensile strength MPa	0.2% pr. stress MPa	% elong.	Electrical conduct. IACS %
CZ102	Red brass	85 Cu, 15 Zn	Soft	310	130	40	37
CZ106	Cartridge brass	70 Cu, 30 Zn	Soft	330	120	70	27
CZ108	Common brass	63 Cu, 37 Zn	Soft	340	130	56	23
			Hard	530	180	5	26
CZ112	Naval brass	62 Cu, 37 Zn, 1 Sn	Soft	370	140	45	26

Table C6 *Properties of tin bronzes*

Alloy	Name	Composition %	Condition	Tensile strength MPa	0.2% pr. stress MPa	% elong.	Electrical conductivity IACS %
Wrought alloys							
PB101	3% phosphor bronze	97 Cu, 3 Sn, 0.02–0.40 P	Soft	320	110	55	15–25
			Hard	580	450	8	
PB103	7% phosphor bronze	95 Cu, 7 Sn, 0.01–0.4 P	Soft	370	130	65	11–15
			Hard	650	570	14	
Cast gunmetals							
G1	Admiralty gunmetal	88 Cu, 10 Sn, 2 Zn	Sand cast	270–340	130–160	13–25	10–11
			Chill cast	250–310	130–170	3–8	
LG2	Leaded gunmetal, eighty-five-three fives	85 Cu, 7 Sn, 3 Zn, 3 Pb	Sand cast	200–270	100–130	13–25	10–15
			Chill cast	200–280	110–140	6–15	

Table C7 *Properties of aluminium bronzes*

Alloy	Name	Composition %	Condition	Tensile strength MPa	0.2% pr. stress MPa	% elong.	Electr. conduct. IACS %
Wrought alloys							
CA101	5% aluminium bronze	95 Cu, 5 Al	Soft	370	140	65	15–18
			Hard	650	540	15	
CA102	8% aluminium bronze	92 Cu, 8 Al	Soft	420	90	50	13–15
			Hard	540	230	10	
Cast alloys							
AB1	Aluminium bronze	88 Cu, 9.5 Al, 2.5 Fe	Sand cast	500–590	170–200	40–18	8–12
			Die cast	540–620	200–270		
AB2	Aluminium bronze	80.5 Cu, 9.5 Al, 5 Fe, 5 Ni	Sand cast	640–700	250–300	20–13	6–8
			Die cast	650–700	250–310		

Table C8 *Properties of a beryllium bronze*

Alloy	Composition %	Condition	Tensile strength MPa	0.2% pr. stress MPa	% elong.	Electr. conduct. IACS %
CB101	98 Cu, 1.7 Be,	Sol. treated	480–500	185–190	45–50	16–78
	0.2–0.6 Co + Ni	Sol. treated + precipitation treated	1150–1160	930–940	5	22–32

Table C9 *Properties of a silicon bronze*

Alloy	Composition %	Condition	Tensile strength MPa	0.2% pr. stress MPa	% elong.	Electr. cond. IACS %
CS101	2.7–3.5 Si, 0.7–1.5 Mn, rem. Cu	Soft sheet	370–390	120–140	60	7
		Hard sheet	600–650	450–490	15–12	6

Table C10 *Properties of cupronickels and nickel silvers*

Alloy	Name	Composition %	Condition	Tensile strength MPa	0.2% proof stress MPa	% elong.	Electr. conduct. IACS %
Cupronickels							
CN101	95/5	94 Cu, 5 Ni, 1 Fe, 0.6 Mn	Soft	280	90	40	
			Cold worked	320–380	300–350	14–10	14
CN105	75/25	75 Cu, 25 Ni, 0.5 Mn	Soft	360	140	40	
			Cold rolled	450–590	390–530	15–3	5
Nickel silvers							
NS103	10%	63 Cu, 27 Zn, 10 Ni	Soft	350–370	100–140	45–65	
			Hard	470–540	430–460	10–12	8
NS105	15%	64 Cu, 21 Zn, 15 Ni	Soft	390–420	130–180	45–55	
			Hard	490–710	370–600	2–15	7

Gold Gold is very ductile and readily cold worked. It has good electrical and thermal conductivity.

Iron The term ferrous alloys is used for the alloys of iron. These alloys include carbon steels, cast irons, alloy steels and stainless steels. Steels have 0.05 to 2% carbon, cast irons 2 to 4.3% carbon. The term carbon steel is used for those steels in which essentially just iron and carbon are present (Table C11). Steels with between 0.10 and 0.25% are termed mild steels, between 0.20 and 0.50% medium-carbon steels and 0.50 to 2% carbon as high-carbon steels. With such steels in the annealed state the tensile strength increases from about 250 MPa at low-carbon content to 900 MPa at high-carbon content, the higher the carbon content the more brittle the alloy. Mild steels are general-purpose steels and used for such applications as joists in buildings, bodywork for cars and ships, screws, nails and wire. Medium-carbon steels are used for shafts and parts in car transmissions, suspensions and steering. High-carbon steels are used for machine tools, saws, axes, hammers, punches and dies. The term low alloy steel is used for alloy steels when the alloying additions are less than 2%, medium alloy between 2 and 10% and high alloy when over 10%. In all cases the carbon content is less that 1%. Example of low alloy steels are manganese steels (Table C12) with strengths of the order of 500 MPa in the annealed state and 700 MPa when quenched and tempered. They are used in applications where higher strengths are required than are possible with carbon steels, e.g. axles and shafts. Stainless steels are high alloy steels with more than 12% chromium and are used where high resistance to corrosion is required. The modulus of elasticity of steels tend to be about 200 to 207 GPa. Cast irons (Tables C13 and C14) typically have tensile strengths of the order of 150 to 500 MPa, a tensile modulus of about 100 to 170 GPa and percentage elongations that are often very low. They can be hard and brittle, being used for manhole covers, heavy-duty piping and machine tool beds.

Table C11 *Properties of plain carbon steels*

% carbon	Condition	Tensile strength MPa	Percentage elongation	Brinell hardness	Izod impact strength J
0.2	Annealed	400	37	115	123
0.4	Annealed	520	30	150	44
	Quench, temper 200°C	910	16	260	
	Quench, temper 430°C	850	21	240	
0.6	Annealed	635	23	180	11
	Quench, temper 200°C	1120	13	320	
	Quench, temper 430°C	1090	14	310	
0.8	Annealed	620	25	170	6
	Quench, temper 200°C	1330	12	390	
	Quench, temper 430°C	1300	13	375	
1.0	Annealed	660	13	190	3
	Normalised	1030	10	290	5
	Quench, temper 200°C	1300	10	400	
	Quench, temper 430°C	1200	12	360	

Table C12 *Properties of manganese steels*

BS reference	Composition %	Condition	Tensile strength MPa	Yield stress MPa	Percentage elongation
150M19	0.2 C, 1.3–1.7 Mn	OQ 860–900°C T 550–660°C	550–700	360	13
150M36	0.32–0.4 C, 1.3–1.7 Mn	OQ 840–870°C, T 550–660°C	625–775	400	18

OQ = oil quench, T = temper

Table C13 *Properties of grey cast irons*

Grade	% carbon equivalent	Tensile strength MPa	% elongation	Brinell hardness	Impact strength* J
150	4.5	150	0.6	100–170	8–13
250	3.85	250	0.5	145–220	13–23
350	3.5	350	0.5	185–260	24–47

* Unnotched 20 mm diameter Izod test pieces

Table C14 *Properties of blackheart, whiteheart and pearlitic malleable cast irons*

BS grade	Min. tensile strength MPa	Min. 0.2% proof stress	Min. % elongation
Blackheart			
B32-10	320	190	10
B35-12	350	200	12
Whiteheart			
W38-12	400	210	8
W40-05	420	230	4
Pearlitic			
P45-06	450	270	6
P55-04	550	340	4
P65-02	650	430	2

Lead Other that its use in lead storage batteries, it finds a use in lead–tin alloys as a metal solder and in steels to improve the machinability.

Magnesium Magnesium is used in engineering alloyed mainly with aluminium, zinc, and manganese (Table C15). The alloys have a very low density and though tensile strengths are only of the order of 250 MPa there is a high strength to weight ratio. The alloys have a low tensile modulus, about 40 GPa. They have good machinability. They find uses in such applications as instrument casings, power tool and electric motor components, car wheels and in the aircraft industry where weight is an important consideration.

Table C15 *Properties of commonly used magnesium alloys*

BS code	Composition %	Condition	0.2% proof stress MPa	Tensile strength MPa	Percentage elongation
Casting alloys					
A8	8 Al, 0.5 Zn, 0.3 Mn	As sand cast	80	140	3
AZ91	9.5 Al, 0.5 Zn, 0.3 Mn	As sand cast	95	135	2
		As die cast	100	170	2
Wrought alloys					
AZ31	3 Al, 1 Zn, 0.3 Mn	Sheet, soft	120	240	16
		Cold worked	160	250	4
		As extruded	130	230	4
		As forged	105	200	7
AZM	6.5 Al, 1 Zn, 0.3 Mn	As extruded	180	260	7
		As forged	160	275	7

Molybdenum Molybdenum has a high density, high electrical and thermal conductivity and low thermal expansivity. At high temperatures it oxides. It is used for electrodes and support members in electronic tubes and light bulbs, and heating elements for furnaces. Molybdenum is however more widely used as an alloying element in steels. In tool steels it improves hardness, in stainless steels it improves corrosion resistance, and in general in steels it improves strength, toughness and wear resistance.

Nickel Nickel is used as the base metal for a number of alloys with excellent corrosion resistance and strength at high temperatures. The alloys are basically nickel–copper and nickel–chromium–iron. The alloys have tensile strengths between about 350 and 1400 MPa, the tensile modulus being about 220 GPa (Tables C16 and C17). They are used for pipes and containers in the chemical industry where high resistance to corrosive atmospheres is required, food-processing equipment and applications, such as gas turbine blades and parts, where strength at high temperatures is required.

Table C16 *Typical properties of Monels at 20°C*

Alloy	Composition %	Condition	Tensile strength MPa	0.2% pr. stress MPa	% elong.
Monel 400, NA 13	Ni 66.5, Cu 31, Fe 1.5, Mn 1.0	Cold worked + annealed	480	170	35
		Cold worked + stress relief	600	400	
Monel K-500, NA 18	Ni 66.5, Cu 30, Al 2.8, Ti 0.5	Cold worked + solution tr. + precipitation treated	900	620	20
		Hot worked + solution tr. + precipitation treated	900	600	

Table C17 *Properties of nickel-base heat-resisting alloys*

Alloy	Composition %	Condition	Temp. °C	Tensile strength MPa	0.2% proof stress MPa	% elong.
Wrought alloys						
Inconel 600, NA 14	Ni 76, Cr 15.5, Fe 8	Hot rolled	20	590	250	50
			400	560	185	50
			600	530	150	40
Nimonic 80A, NA 20, HR 1	Ni 73, Cr 19.5, Co 1, Ti 2.25, Al 1.4, Fe 1.5	Hot rolled, solution tr. and precipitation	20	1240	740	24
			400	1150	680	26
			600	1080	620	20
			800	620	490	24
Incoloy 825, NA 16	42 Ni, 30 Fe, 21 Cr, 2.3 Cu, 0.09 Ti, 3 Mo	Hot worked + solution treated	20	590	220	30
			700	335	210	49
			1000	65	35	75
Cast alloys						
Nimocast 80, ANC 9	70 Ni, 20 Cr, 0.6 Si, 2.0 Co, 1.2 Al, 2.6 Ti, 0.6 Mn, 0.08 C	As cast	700	970	680	12
			1000	75	40	82

Niobium It has a high melting point, good oxidation resistance and low modulus of elasticity. Niobium alloys are used for high temperature items in turbines and missiles. It is used as an alloying element in steels.

Palladium This metal is highly resistant to corrosion. It is alloyed with gold, silver or copper to give metals which are used mainly for electrical contacts.

Platinum The metal has a high resistance to corrosion, is very ductile and malleable, but expensive. It is widely used in jewellery. Alloyed with elements such as iridium and rhodium, the metal is used in instruments for items requiring a high resistance to corrosion.

Silver Silver has a high thermal and electrical conductivity, and is very soft and ductile.

Tantalum Tantalum is high melting point, highly acid-resistant, very ductile metal. Tantalum-tungsten alloys have high melting points, high corrosion resistance and high tensile strength.

Tin Tin has a low tensile strength, is fairly soft and can be very easily cut. Tin plate is steel plate coated with tin, the tin conferring good corrosion resistance. Solders are essentially tin alloyed with lead and sometimes antimony. Tin alloyed with copper and antimony gives a material widely used for bearing surfaces.

Titanium Titanium as commercially pure or alloy has a high strength coupled with a relatively low density. It retains its properties over a wide temperature range and has excellent corrosion resistance. Tensile strengths are typically of the order of 1100 MPa and tensile modulus about 110 GPa (Table C18). Titanium and its alloys are used for jet engine parts and for marine and chemical plant parts.

Table C18 *Composition and properties of commercially pure annealed titanium*

Composition %	Temperature °C	Tensile strength MPa	Yield stress MPa	% elongation
99.5 Ti	20	330	240	30
	300	150	95	32
99.2 Ti	20	440	350	28
	300	220	120	35
99.1 Ti	20	520	450	25
	300	240	140	34
99.0 Ti	20	670	590	20
	300	310	170	37

Tungsten This is a dense metal with the highest melting point of any metal (3410°C). It is used for light bulb and electronic tube filaments, electrical contacts, and as an alloying elements in steels. As whiskers it is used in many metal-whisker composites.

Zinc Zinc has very good corrosion resistance and hence finds a use as a coating for steel, the product being called galvanised steel. It has a low melting point and hence zinc alloys are used for products such as small toys, cogs, shafts, door handles, etc. produced by die casting. Zinc alloys are generally about 96% zinc with 4% aluminium and small amounts of other elements or 95% zinc with 4% aluminium, 1% copper and small amount of other elements. Such alloys have tensile strength of about 300 MPa, elongations of about 7–10% and a Brinell hardness of about 90 (Table C19).

Table C19 *Properties of zinc pressure die casting alloys*

Alloy	Condition	Tensile strength MPa	% elongation	Impact strength J
MAZAK 3	As cast	286	15	57
	Stabilised	273	17	61
MAZAK 5	As cast	335	9	58
	Stabilised	312	10	60

Engineering polymers

The following is an alphabetical listing of the main polymers used in engineering, together with brief notes of their main characteristics. Table C20 gives typical properties of a range of thermoplastics, Table C21 the properties of a range of thermosets and Table C22 the properties of a range of elastomers.

Table C20 *Properties of thermoplastics commonly used in engineering*

Polymer	State	T_g °C	Max. temp. °C	Tensile strength MPa	Tensile modulus GPa	% elongation	Impact property
Polyethylene							
High density	SC	−120	125	22–38	0.4–1.3	50–800	2
Low density	SC	−90	85	8–16	0.1–0.3	100–600	3
Polypropylene	SC	−10	150	30–40	1.1–1.6	50–600	2
Polyvinyl chloride							
With no plasticiser	G	87	70	52–58	2.4–4.1	2–40	2
With low plasticiser	G		100	28–42		200–250	2
Polystyrene							
No additives	G	100	70	35–60	2.5–4.1	2–40	1
Toughened	G		70	17–24	1.8–3.1	8–50	2
ABS	G	100	70	17–58	1.4–3.1	10–140	2–3
Polycarbonate	G	150	120	55–65	2.1–2.4	60–100	2–3
Acrylic	G	100	100	50–70	2.7–3.5	5–8	1
Polyamides							
Nylon 6	SC	50	110	75	1.1–3.1	60–320	2–3
Nylon 6.6	SC	55	110	80	2.8–3.3	60–300	2–3
Nylon 11	SC	46	110	50	0.6–1.5	70–300	2–3
Polyethylene terephthalate	SC	69	120	50–70	2.1–4.4	60–100	2
Polyacetals, homopolymer	SC	−76	100	60	3.6	15–75	2
PTFE	SC	−120	260	14–35	0.4	200–600	3
Cellulose acetate	G	120	70	24–65	1.0–2.0	5–55	2
Cellulose acetate butyrate	G	120	70	17–20	1.0–2.0	8–80	2

State: SC is semi-crystalline, G is glass. Impact property: 1 is brittle, 2 is tough when unnotched but brittle when bluntly notched, 3 is tough under all conditions, all referring to about 20°C.

Table C21 *Properties of thermosets*

Polymer	Density Mg/m³	Tensile strength MPa	Tensile modulus GPa	Percentage elongation	Max. service temp. °C
Phenol formaldehyde					
Unfilled	1.25–1.30	35–55	5.2–7.0	1.0–1.5	120
Wood flour filler	1.32–1.45	40–55	5.5–8.0	0.5–1	150
Urea formaldehyde					
Cellulose filler	1.5–1.6	50–80	7.0–13.5	0.5–1	80
Wood flour filler	1.5–1.6	40–55	7.0–10	0.5–1	
Melamine formaldehyde					
Cellulose filler	1.5–1.6	55–85	7.0–10.5	0.5–1	95
Glass filled	2.0	35–138	12–17	0.5	
Epoxy resin					
Cast	1.15	60–100	3.2		
60% glass fabric	1.8	200–240	21–25		200
Polyester					
Unfilled	1.3	55	2.4		200
30% glass fibre	1.5	120	7.7	3	

Table C22 *Properties of elastomers*

Elastomer	Tensile strength MPa	Percentage elongation	T_g °C	Service temps. °C	Resistance to oil and greases	Resilience
Natural rubber	20	800	–73	–50 to +80	Poor	Good
Butadiene-styrene	24	600	–58	–50 to +80	Poor	Good
Butadiene-acrilonitrile	28	700	–55	–50 to +100	Excellent	Fair
Butyl	20	900	–79	–50 to +100	Poor	Fair
Polychloroprene	25	1000	–48	–50 to +100	Good	Good
Ethyl-propylene	20	300	–75	–50 to +100	Poor	Good
Polypropylene oxide	14	300		–20 to +170	Poor	
Fluorosilicone	8	300	–123	–100 to +200	Good	Fair
Polysulphide	9	500	–50	–50 to + 80	Good	Fair
Polyurethane	40	650	–60	–55 to + 80	Poor	Poor
Styrene-butadiene-styrene	14	700		–60 to +80	Poor	

Resilience is the capacity of a material to absorb energy in the elastic range, being measured from the area under the elastic portion of the stress–strain graph. To illustrate this property; a high resilience rubber ball will bounce to a much higher height than a low resilience one.

Acetals Acetals, i.e. polyacetals, are thermoplastics with properties and applications similar to those of nylons. A high tensile strength (70 MPa) is retained in a wide range of environments. They have a high tensile modulus (3.6 GPa) and hence stiffness, high impact resistance and a low coefficient of friction. Ultraviolet radiation causes surface damage. They are used for such applications as pipe fittings, parts for water pumps and washing machines, car instrument housings, bearings and hinges.

Acrylics Acrylics are transparent thermoplastics, trade names for such materials including Perspex and Plexiglass. They have high tensile strength (50 to 70 MPa) and tensile modulus (2.7 to 3.5 GPa), hence stiffness, good impact resistànce and chemical resistance, but a large thermal expansivity. They are used for light fittings, lenses for car lights, signs and shower cabinets.

Acrylonitrile butadiene styrene (ABS) ABS is a thermoplastic polymer giving a range of opaque materials with good impact resistance, ductility and moderate tensile (17 to 58 MPa) and compressive strength. It has a reasonable tensile modulus (1.4 to 3.1 GPa) and hence stiffness, with good chemical resistance. It is used as casings for telephones, vacuum cleaners, hair dryers, TV sets and radios.

Butadiene acrylonitrile This is an elastomer, generally referred to as nitrile or Buna-N rubber (NBR). It has excellent resistance to fuels and oils and is used for gaskets, hoses, seals and rollers.

Butadiene styrene This is an elastomer and is very widely used as a replacement for natural rubber because of its cheapness. It has good wear and weather resistance, good tensile strength, but poor resilience, poor fatigue strength and low resistance to fuels and oils. It is used in the manufacture of car tyres, hose pipes and conveyor belts.

Butyl Butyl, i.e. isobutene-isoprene copolymer, is an elastomer. It is extremely impermeable to gases and is used for the inner linings of tubeless tyres, steam hoses and diaphragms.

Cellulosics This term encompasses cellulose acetate, cellulose acetate butyrate, cellulose acetate propionate, cellulose nitrate and ethyl cellulose. All are thermoplastics. Cellulose acetate is a transparent material. Additives are required to improve toughness and heat resistance. Cellulose acetate butyrate is similar to cellulose acetate but less temperature sensitive and with a greater impact strength. it has a tensile strength of 18 to 48 MPa and a tensile modulus of 0.5 to 1.4 GPa. Cellulose nitrate colours and becomes brittle on exposure to sunlight. It also burns rapidly. Ethyl cellulose it tough and has low flammability. Cellulosics are used for spectacle frames, tool handles, toys, containers, cable insulation and lenses for instrument panel lights.

Chlorosulphonated polyethylene This is an elastomer, trade name Hypalon. It has excellent resistance to ozone with good chemical resistance, fatigue and impact properties. It is used for flexible hose for oil and chemicals, tank linings, cable insulation and shoe soles.

Epoxies Epoxy resins are, when cured, thermosets. They are frequently used with glass fibres to form composites. Such composites have strength of the order of 200 to 420 MPa and stiffness about 21 to 25 GPa.

Ethylene propylene This is an elastomer. The copolymer form, EPM, and the terpolymer form, EPDM, have very high resistance to oxygen, ozone and heat. It is used for electrical insulation, hoses and belts.

Ethylene vinyl acetate This is an elastomer which has good flexibility, impact strength and electrical insulation properties. It is used for cable insulation, flexible tubing and gaskets.

Fluorocarbons These are polymers consisting of fluorine attached to carbon chains. See Polytetrafluoroethylene.

Fluorosilicones See Silicone rubbers.

Melamine formaldehyde The resin, a thermoset, is widely used for impregnating paper to form decorative panels, and as a laminate for table and kitchen unit surfaces. It is also used with fillers for moulding knobs, handles, etc. It has good chemical and water resistance, good colourability and good mechanical strength (55 to 85 MPa) and stiffness (7.0 to 10.5 GPa).

Natural rubber This is an elastomer. It is inferior to synthetic rubbers in oil and solvent resistance and oxidation resistance. It is attacked by ozone. It is used for tyres, hose and gaskets.

Nylons The term nylon is used for a range of thermoplastic materials having the chemical name of polyamides. A numbering system is used to distinguish between the various forms, the most common engineering ones being nylon 6, nylon 6.6 and nylon 11. Nylons are translucent materials with high tensile strength and of medium stiffness. Tensile strengths are typically about 75 MPa and the tensile modulus about 1.1 to 3.3 GPa. Additives such as glass fibres are used to increase strength. Nylons have low coefficients of friction, which can be further reduced by suitable additives. For this reason they are widely used for gears, rollers, bearings and bushes. They are also used for housings for power tools, electric plugs and sockets and as fibres in clothing. All nylons absorb water.

Phenol formaldehyde This is a thermoset and is mainly used as a reinforced moulding powder. It is low cost, and has good heat resistance, dimensional stability, and water resistance. Unfilled it has a tensile strength of 35 to 55 MPa and a tensile modulus of 5.2 to 7.0 GPa. It is used for electrical plugs and sockets, switches, door knobs and handles.

Polyacetal See Acetals.

Polyamides See Nylons.

Polycarbonates Polycarbonates are transparent thermoplastics with high impact strength, high tensile strength (55 to 65 MPa), high dimensional stability and good chemical resistance. They are moderately stiff (2.1 to 2.4 GPa). They have good heat resistance and can be used at temperatures up to 120°C. They are used for street lamp covers, infant feeding bottles, machine housings, safety helmets, cups and saucers.

Polychloroprene This, usually called neoprene, is an elastomer. It has good resistance to oils and good weathering resistance. It is used for oil and petrol hoses, gaskets, seals, diaphragms and chemical tank linings.

Polyesters Two forms are possible, thermoplastics and thermosets. Thermoplastic polyesters have good dimensional stability, excellent electrical resistivity and are tough. They discolour when subject to ultraviolet radiation. Thermoset polyesters are generally used with glass fibres to form composite materials that are used for boat hulls, building panels and stackable chairs.

Polyethylene Polyethylene, or polythene, is a thermoplastic material. There are two main types: low density (LDPE) which has a branched polymer chain structure and high density (HDPE) with linear chains. Materials composed of blends of the two forms are available. LDPE has a fairly low tensile strength (8 to 16 MPa) and tensile modulus (0.1 to 0.3 GPa), with HDPE being stronger (22 to 38 MPa) and stiffer (0.4 to 1.3 GPa). Both forms have good impermeability to gases and very low absorption rates for water. LDPE is used for bags, squeeze bottles, ball-point pen tubing, and wire and cable insulation. HDPE is used for piping, toys and household ware.

Polyethylene terephthalate (PET) This is a thermoplastic polyester. It has good strength (50 to 70 MPa) and stiffness (2.1 to 4.4 GPa), is transparent and has good impermeability to gases. It is widely used as bottles for fizzy drinks. It is also used for electrical plugs and sockets, recording tape and wire insulation.

Polypropylene Polypropylene is a thermoplastic material with a low density, reasonable tensile strength (30 to 40 MPa) and stiffness (1.1 to 1.6 GPa). Its properties are similar to those of polyethylene. Additives are used to modify the properties. It is used for crates, containers, fans, car fascia panels, radio and TV cabinets, toys and chair shells.

Polypropylene oxide This is an elastomer with excellent impact and tear strengths, good resilience and good mechanical properties. It is used for electrical insulation.

Polystyrene Polystyrene is a transparent thermoplastic. It has moderate tensile strength (35 to 60 MPa), reasonable stiffness (2.5 to 4.1 GPa), but is fairly brittle and exposure to sunlight results in yellowing. It is attacked by many solvents. Toughened grades, produced by blending with rubber, have better impact properties. They have a strength of about 17 to 42 MPa and stiffness of 1.8 to 3.1 GPa. This form is used for vending machine cups, casings for cameras, radios and TV sets. Foamed, or expanded as it is generally termed, polystyrene is used for insulation and packaging.

Polysulphide This is an elastomer with excellent resistance to oils and solvents, and low permeability to gases. It can however be attacked by micro-organisms. It is used for cable covering, coated fabrics and sealants in building work.

Polysulphone This is a strong, comparatively stiff, thermoplastic which can be used to a comparatively high temperature. It has good dimensional stability and low creep. It has a strength of about 70 MPa and a stiffness of about 2.5 GPa. It burns with difficulty and does not present a smoke hazard; it thus finds uses in aircraft as parts on passenger service units, circuit boards and cooker control knobs.

Polytetrafluoroethylene (PTFE) PTFE is a tough and flexible thermoplastic which can be used over a very wide temperature range. Because other materials will not bond with it, the material is used as a coating to items where non-stick facilities are required, e.g. non-stick domestic cooking pans.

Polyvinyls Polyvinyls are thermoplastics and include polyvinyl acetate, polyvinyl butyral, polyvinyl chloride (PVC), chlorinated polyvinyl chloride and vinyl copolymers. Polyvinyl acetate (PVA) is widely used in adhesives and paints. Polyvinyl butyral (PVB) is mainly used as a coating material or adhesive. PVC has high strength (52 to 58 MPa) and stiffness (2.4 to 3.1 GPa), being a rigid material. It is frequently combined with plasticisers to give a lower strength, less rigid, material. Without plasticiser it us used as piping for waste and soil drainage systems, rain water pipes, lighting fittings and curtain rails. With plasticiser it is used for plastic raincoats, bottles, shoe soles, garden hose pipes and inflatable toys. Chlorinated PVC is hard and rigid with excellent chemical and heat resistance. Vinyl copolymers can give a range of properties according to the constituents and their ratio. A common copolymer is vinyl chloride with vinyl acetate in the ratio 85 to 15. This is a rigid material. A more flexible form has the ratio 95 to 5.

Silicone rubbers Silicone rubbers or, as they are frequently called, fluorosilicone rubbers have good resistance to oils, fuels and solvents at high and low temperatures. They do however have poor abrasion resistance. They are used for electric insulation seals and shock mounts.

Styrene-butadiene-styrene This is called a thermoplastic rubber. Its properties are controlled by the ratio of styrene to butadiene. The properties are comparable to those of natural rubber. It is used for footwear, carpet backing and in adhesives.

Urea formaldehyde This is a thermosetting material and has similar applications to melamine formaldehyde. Surface hardness is very good. The resin is also used as an adhesive.

Engineering ceramics

The term ceramics covers a wide range of materials and here only a few of the more commonly used engineering ceramics are considered.

Alumina Alumina, i.e. aluminium oxide, is a ceramic which finds a wide variety of uses (Table C23). It has excellent electrical insulation properties and resistance to hostile environments. Combined with silica it is used as refractory bricks.

Table C23 *Properties of alumina-based ceramics*

Material	Density Mg/m^3	Coeff. of expansion 10^{-6} K^{-1}*	Elastic modulus GPa	Short-term strength MPa	Max. temp. °C	Resist-ivity Ω m	Dielectric constant at 50 Hz*	Dielectric strength kV/mm
Alumina 99%	3.7	5.9	380	**	1600	10^{12}	9	17
Alumina 95%	3.5	5.6–5.9	320	**	1400	10^{12}	9	15
Alumina 90%	3.4	5.0–6.0	260	**	1200	10^{12}	9	15

* 25°C, ** depends on grain size, 200–400 MPa for fine grains, 100–200 MPa medium to coarse grains

Boron Boron fibres are used as reinforcement in composites with materials such as nickel.

Boron nitride This ceramic is used as an electric insulator.

Carbides A major use of ceramics is, when bonded with a metal binder to form a composite material, as cemented tips for tools. These are generally referred to as bonded carbides, the ceramics used being generally carbides of chromium, tantalum, titanium and tungsten.

Cement There are several types of cement. Typically, Portland cement consists of 60 to 64% calcium oxide, 19 to 25% silicon oxide, 5 to 9% aluminium oxide, 2 to 4% iron oxide. When water is mixed with cement, a reaction occurs which results in a silicate structure being formed. Portland cement is used as the binder/matrix in the manufacture of concrete, a composite material involving cement, gravel and sand.

Chromium carbide See Carbides.

Chromium oxide This ceramic is used as a wear resistant coating.

Clay products Many ceramic materials are made primarily from clay to which other materials such as silica and feldspar have been added. The materials are mixed with water and the product formed. It is then dried and fired to produce the ceramic bonds. Earthenware, such as drain pipes, are fired at a low temperature and have a relatively porous structure. China and porcelain have high firing temperatures. This results in some of the mixture being converted into a clear glass and so a translucent material.

Glasses The basic ingredient of most glasses is silica, a ceramic. Glasses (Table C24) tend to have low ductility, a tensile strength which is markedly affected by microscopic defects and surface scratches, low thermal expansivity and conductivity (and hence poor resistance to thermal shock), good resistance to chemicals and good electrical insulation properties. Glass fibres are frequently used in composites with polymeric materials.

Table C24 *Properties of glasses*

Glass	Composition %	Density Mg/m^3	Coeff. of expansion $10^{-6} K^{-1}$	Elastic modulus GPa	Hardness HK0.1	Max. temp. °C
Soda lime glass	70–75 SiO_2, 0.5–3 Al_2O_3, 12–17 Na_2O, 4.5–12 CaO, 0–3 MgO	2.5	9.2	70	460	460
Borosilicate glass	60–80 SiO_2, 1–4 Al_2O_3, 10–12 B_2O_3	2.2	3.2	67	420	490

Note: the hardness scale HK0.1 is the Knoop test with a 100 g load

Kaolinite This ceramic is a mixture of aluminium and silicon oxides, being a clay. Large electrical insulators, e.g. those used with overhead high voltage cables, are made by firing a mixture of kaolinite, feldspar and silica.

Magnesia Magnesia, i.e. magnesium oxide, is a ceramic and is used to produce a brick called a dolomite refractory.

Pyrex This is a heat resistant glass, being made with silica, limestone and boric oxide. See Glasses.

Silica Silica forms the basis of a large variety of ceramics. It is, for example, combined with alumina to form refractory bricks and with magnesium ions to form asbestos. It is the basis of most glasses.

Silicon nitride This ceramic is used as the fibre in reinforced materials, such as epoxies.

Soda glass This is the common window glass, being made from a mixture of silica, limestone and soda ash. See Glasses.

Tantalum carbide See Carbides.

Titanium carbide See Carbides.

Tungsten carbide See Carbides.

Composites

Ceramic matrix Typical ceramic matrix materials are alumina with silicon carbide whiskers (25% silicon carbide: elastic modulus 340 GPa and strength 900 MPa) and Pyrex glass with aluminium oxide fibres (40% fibres: strength about 300 MPa). There is a marked improvement in the toughness of the material when compared with the ceramic matrix material alone. Industrial cutting tools are made with alumina reinforced with silicon carbide whiskers.

Laminates These are made by stacking together sheets of materials that generally have unidirectional properties. When stacked cross-ply, i.e. the fibre orientations in the sheets at right angles to each other, the result can be a product that has the same properties in the two right-angled directions. An example of such a material is plywood with thin layers, plies, of wood being laminated together with an adhesive. Three-ply wood consists of a first layer with fibres in one direction, a second layer with fibres at right angles, and then a final layer with fibres in the same direction as the initial layer. Using plies with a tensile modulus of 16 GPa parallel to the fibres and 1.1 GPa at right angles, three-ply has tensile modulus of 12 GPa parallel to the fibres in the outer layers and 0.9 GPa at right angles. With nine-plies this becomes about 11 GPa and 3 GPa.

Metal matrix Alumina and silicon carbide have been used as reinforcement for metals. An example is titanium with silicon carbide fibres (35% silicon carbide: tensile modulus 210 GPa and tensile strength 1700 MPa). The composite is stiffer and stronger than the metal alone, but less ductile.

Natural composites Wood is an example of a natural composite. It consists of longitudinal cellulose cells bound together with lignin. The properties along the fibre direction are different from those at right angles to it. For example, the tensile modulus of ash is 16 GPa in the grain direction, 0.9 GPa at right angles to it. Douglas fir is 16 GPa in the grain direction and 0.8 GPa at right angles to it. Bone is another example of a natural composite. Bone, such as the human femur, has a tensile modulus of about 20 GPa and a compressive modulus of about 1 GPa. The compressive strength is about 5 to 20 MPa.

Polymer matrix Typical polymer matrix materials are epoxy with glass fibre (70% glass fibres: tensile modulus 40 GPa and tensile strength 750 MPa in the direction of the fibres), polyester with glass fibre (50% glass fibre: tensile modulus 40 GPa and tensile strength 760 MPa in the direction of the fibres), nylon with carbon fibre (40% carbon fibre: tensile modulus 20 GPa and tensile strength 250 MPa). Such materials are much stiffer and stronger than the polymer alone and have stiffnesses and strengths as good as some metal alloys.

Answers

The following are the numerical answers to problems and brief clues as to the form of the answers for other problems.

Chapter 2

1 These might include (a) ease of forming in one piece, easily cleaned, stain resistance, waterproof; (b) stiff, strong, cheap; (c) leak proof, suitable for hot liquids, cheap, not easily broken; (d) good conductor, flexible; (e) cheap to make, wear resistant during handling, stiff; (f) withstands changing forces, stiff, strong, withstands impact forces; (g) attractive appearance, cheap to form

2 (a) Stainless steel, (b) wood, (c) china (a ceramic), (d) copper, (e) alloys of copper (cupronickel or bronze depending on the colour of the coins), (f) steel, (g) plastic, e.g. ABS

3 (a) Modulus, (b) ductility, percentage elongation, (c) fracture toughness, (d) strength, (e) electrical resistivity/conductivity, (f) thermal conductivity, (g) corrosive properties

4 Strong and brittle

5 Strong and tough

6 20 MPa

7 0.67%

8 50 kN

9 12%

10 50 kN

11 The bronze is stronger and more ductile

12 Stronger in compression, brittle

13 Ductile above 0°C, brittle below

14 Reasonably good

15 Very low resistivity, of the order of 10^{-8} Ω m

16 0.0125 Ω

17 Thermoplastics: flexible, soft, can be formed by heating; thermosets: rigid, hard, cannot be formed by heating

18 Brittle, must not be subject to sudden forces or sudden changes in temperature

19 128 MPa/Mg m^{-3}, 33 MPa/Mg m^{-3}, 0.78 £/MPa, 6 £/MPa

Chapter 3

1 (a) 420 MPa, (b) 62%, (c) 18–40%, (d) 355 MPa, (e) 510 MPa, (f) 1020–1070 kg/m^3, (g) 3.0–4.5 MPa $m^{-1/2}$, (h) 20 MPa, (i) 11–13 × $10^{-2}K^{-1}$, (j) 1.4–3.1 GPa

2 Cast iron 0.014 GPa/kg m^{-3}, Al alloy 0.027 GPa/kg m^{-3}, PVC 0.002 GPa/kg m^{-3}
3 Steel 220 GPa, Al alloy 71 GPa, Polypropylene 1–2 GPa, Composite 20 GPa
4 120 MPa/Mg m^{-3}
5 1.4–3.1 GPa, in the high range of modulus values for plastics
6 470–570 MPa, 170–280 MPa, 18–35%
7 150M36: 620–770 MPa, 400 MPa, 18%; 530M40: 700–850 MPa, 525 MPa, 17%
8 LM6
9 Polyacetal
10 220M44
11 Electrolytic or oxygen-free copper
12 Aluminium alloys

Chapter 4

1 (a) 61 GPa, (b) 380 MPa
2 (a) 10.8 MPa, (b) 1.1 GPa
3 (a) 660 MPa, (b) 425 MPa, (c) 200 GPa
4 31.1%
5 (a) 480 MPa, (b) 167 GPa
6 300 MPa, 280 MPa
7 2.5 GPa, 80 MPa
8 See Figure A1
9 Stronger and less ductile
10 (a) Titanium alloy, (b) nickel alloy
11 Cellulose acetate
12 Becoming more ductile
13 Becoming more brittle
14 As the temperature drops becoming more brittle
15 Becoming more ductile
16 HV 198
17 HV 275
18 HV 71
19 HB 217
20 HB 57
21 After exposure breakdown voltage decreases
22 Ni-Cr alloy most corrosion resistant
23 Increasing carbon reduces oxidation, increasing chromium reduces oxidation
24 Industrial pollutants more damaging than marine conditions, with rural surroundings being least corrosive
25 (a) Hardness test, (b) impact test, (c) tensile test for the modulus of elasticity, (d) impact, or tensile or hardness, test, (e) bend test

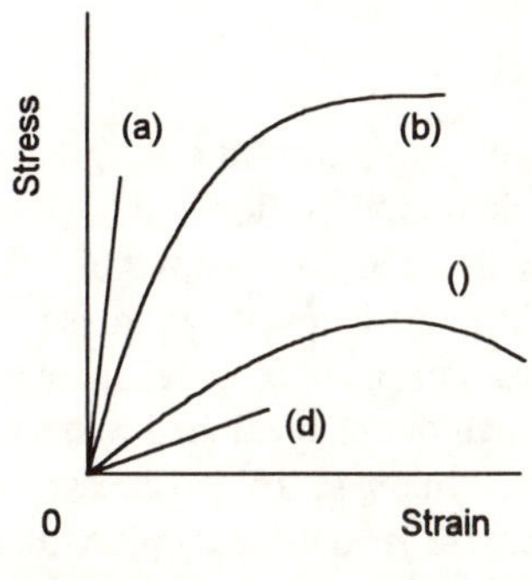

Figure A1 *Chapter 4, Problem 8*

Chapter 5

1 A crystal within a metal, i.e. a region of orderly packing of atoms
2 A mixture of two or more elements, e.g. iron and carbon in steel
3 Ferrous alloys have iron as the main constituent, a non-ferrous alloy a metal other than iron

4 An array of grains, i.e. crystals, within which there are orderly arrays of atoms
5 More ductile the bigger the grains.
6 Elongated grains give different properties in the directions of the grains compared with at right angles
7 Grains become elongated and distorted with an increased number of dislocations; tensile strength and hardness increases, ductility decreases
8 See Figure 5.16
9 (a) Large grain, few dislocations, (b) small grain, many dislocations.
10 Increase in dislocations and hence an increase in yield strength, tensile strength and hardness but a decrease in ductility
11 See Figures 5.29 and 5.30. The greater the crystallinity the greater the density, melting point and strength
12 LDPE is a branched polymer with less crystallisation than HDPE which is a linear polymer. See Table 5.6
13 (a) To protect against UV and resist deterioration, (b) to make more flexible, (c) to reduce cost, increase perhaps strength, impact strength, resistivity, or reduce friction
14 See Table 5.8
15 Makes it more rigid
16 See Section 5.6.2
17 36.4 GPa, in direction of fibres
18 182.2 GPa
19 205 GPa
20 The long fibres give directionality of properties and a greater improvement in strength and modulus than random fibres
21 (a) Recrystallisation and grain growth, ductility improves; (b) martensite forms as carbon atoms become trapped, increase in hardness; (c) some carbon atoms diffuse out of martensite, increase in ductility; (d) fine particles slowly move out of quenched material into dislocations and grain boundaries, increase in hardness; (e) surface layers become martensitic, increase in hardness, (f) carbon diffuses into outer layers, increase in surface hardness
22 Annealing gives grain growth and a soft structure; quenching gives martensitic structure and increase in hardness, strength and brittleness; tempering allows carbon atoms to diffuse out of the martensite and so reduce the structural distortion and hence brittleness
23 Annealing gives grain growth and a soft, weak structure, precipitation hardening causes fine particles to become lodged at grain boundaries and dislocations, hence increasing the strength, hardness and brittleness
24 The hammer head is forged and then the striking surface is surface hardened, possibly by flame hardening followed by tempering
25 The blade has to be tough with the teeth hard. Cast ingots are hot rolled, then blade size strips cut out and teeth machined. Surface hardening, flame hardening, is then used for the teeth followed by tempering in order to achieve the required hardness and not too brittle a state for the teeth

Chapter 6

1 Better surface finish with cold drawing. The heating anneals the material to make it soft and ductile
2 Oriented distorted grains and hence work hardened with directionality of properties
3 Die casting
4 Slow cooling gives high degree of crystallinity
5 Molecules aligned along the direction of the extrusion
6 Molecules aligned with the direction of stretching
7 Gives molecular alignment and improves the strength
8 Annealing gives grain growth and a soft, weak structure; work hardening distorts grains and introduces dislocations with the result that the material is stronger, harder and more brittle
9 See Figure 6.8
10 See Figure 6.8. Above the recrystallisation temperature no work hardening occurs
11 Grains elongated in direction of rolling
12 Cold rolled has distorted grains, is work hardened and directionality of properties. Hot rolled has large grains, is ductile and has surface oxide layers
13 Greater than 400°C, probably about 500°C
14 (a) More distortion and dislocations, increased hardness and brittleness, (b) up to 300°C, (c) above 300°C
15 (a) About 110 HV, (b) about 30 HV, (c) roll, anneal, roll, anneal, roll so that the final rolling gives less than 10% reduction
16 Die pressing or reaction bonding
17 (a) Hand lay-up, (b) sheet moulding
18 (a) Injection moulding, (b) vacuum forming, (c) compression moulding
19 (a) Die casting, (b) sand casting, (c) deep drawing

Chapter 7

1 (a) Aluminium alloy, e.g. LM9, die cast; (b) carbon steel, about 0.3% carbon, or low alloy steel such as a chromium-molybdenum steel, hot rolling; (c) unplasticised PVC, extrusion; (d) high density polyethylene, injection moulding; (e) high density polyethylene, extrusion; (f) nylon, injection moulding or die cast aluminium alloy, e.g. LM9; (g) acrylic (polymethyl methacrylate), injection moulding; (h) medium carbon steel, about 0.4% carbon, or low alloy steel such as 530M40, forged, quenched and tempered; (i) polypropylene, injection moulded or aluminium alloy, e.g. LM6, die cast; (j) ABS, injection moulding
2 (a) Thermoset, e.g. urea formaldehyde, moulding; (b) medium carbon steel or stainless steel, forging; (c) copper, e.g. C106, extrusion; (d) ABS, injection moulding; (e) wood
3 See the British Standards

Index